Richard Berk

Criminal Justice Forecasts of Risk

A Machine Learning Approach

 Springer

Richard Berk
University of Pennsylvania
School of Arts & Sciences
Philadelphia, PA
berkr@sas.upenn.edu

ISSN 2191-5768   e-ISSN 2191-5776
ISBN 978-1-4614-3084-1   e-ISBN 978-1-4614-3085-8
DOI 10.1007/978-1-4614-3085-8
Springer New York Heidelberg Dordrecht London

Library of Congress Control Number: 2012933586

Printed on acid-free paper

Springer is part of Springer Science+Business Media (www.springer.com)

Preface

This book is an effort to put in one place and in accessible form the most recent work on forecasting re-offending by individuals already in criminal justice custody. Much of that work is my own and comes from over two decades of close collaborations with a number of criminal justice agencies. After many requests to provide in one place an account of the procedures I have used, I agreed to write this book. What I hope distinguishes the material from what has come before is the use of machine learning statistical procedures coupled with very large datasets, an explicit introduction of the relative costs of forecasting errors as the forecasts are constructed, and an exclusive statistical focus on maximizing forecasting accuracy. Whether the forecasts that result are "good enough," I leave to the reader.

The audience for the book is graduate students and researchers in the social sciences, and data analysts in criminal justice agencies. Formal mathematics is used only as necessary or in concert with more intuitive explanations. A working knowledge of the generalized linear model is assumed. All of the empirical examples were constructed using the programming language R, in part because most of the key tools are not readily available in the usual social science point-and-click statistical packages. R runs on a wide variety of platforms and is available at no cost. It can be downloaded from www.r-project.org/.

A very large number of individuals have helped me as my criminal justice forecasting activities have evolved. On technical matters, I owe a special debt to David Freedman who put up with my questions over 10 years of collaboration. More recently, I have benefited enormously working the colleagues in Penn's Department of Statistics, especially Larry Brown, Andreas Buja, Ed George, Mikhail Traskin and Linda Zhao. On policy matters my mentor Peter Rossi provided a foundation on which I still draw. My hands-on education has come from a number of criminal justice officials starting with several talented individuals working for the California Department of Corrections in the 1990s: George Lehman, Penny O'Daniel and Maureen Tristan. More recently, Ellen Kurtz of Philadelphia's Adult Department of Probation and Parole was an early supporter and adopter modern forecasting methods from whom I learned a lot. Catherine McVey, chairman of the Pennsylvania Board of Probation and Parole played a key role in educating me about the parole decision process and the *real politik* in which parole decisions are made. Jim Alibirio, as a key IT and data analyst person working for the parole board, was a true data maven and wonderful colleague as we together developed the required forecasting procedures. Other practitioners to whom I am especially indebted are Mark Bergstrom Executive Director of the Pennsylvania Sentencing Commission, and Sarah Hart, a Deputy Prosecutor (and much more) in the Philadelphia district attorney's office. A sincere thanks to one and all.

All of the analyses reported in the book would have been impossible without the data on which they are based. Collecting those data required hard work by a number of individuals from several criminal justice agencies at state and local levels. An important subset of the work was supported by a grant to the State of Pennsylvania from the National Institute of Justice — "Projecting Violent Re-Offending in a Parole Population: Developing A Real-Time Forecasting Procedure to Inform Parole Decision-Making" (# 2009-IJ-CX-0044). My most sincere thanks to NIJ for the funding and to Patrick Clark, who is a knowledgeable and constructive grant monitor with a great sense of humor.

Finally, I am a notoriously bad proof reader and can always use help with the quality of my arguments and prose. Justin Bleich and Anna Kegler carefully read an earlier draft of this manuscript and provided all manner of assistance. However, the responsibility for any remaining typos or unclear prose is surely mine.

Contents

Chapter 1
Getting Started

Abstract This chapter provides a general introduction to forecasting criminal behavior in criminal justice settings. Criminal justice forecasting is a tool that has been used by decision-makers since at least the 1920s. Over time, statistical methods have replaced clinical methods, leading to improvements in forecasting accuracy. The gains were at best gradual until recently, when the increasing availability of very large datasets, powerful computers, and new statistical procedures began to produce dramatic improvements. It is important to note that criminal justice forecasting is inextricably linked to stakeholder decision-making. As such, there are always political considerations, ethical complexities, and judgement calls for which there can be no technical fix.

1.1 Setting the Stage

All forecasts use information on hand to help anticipate outcomes not yet observed. Ideally, it is a three-step process. In the first step, relationships between predictive information and one or more outcomes are documented or assumed. In the second step, forecasting accuracy is evaluated. If the accuracy is sufficient, there is a third step: when subsequently there is predictive information, but the outcomes are unknown, the relationships established earlier permit projections to what the unknown outcome might be.

This is a book about forecasting whether an individual already in the custody of the criminal justice system will subsequently re-offend. Perhaps the most familiar example is forecasts used by parole boards to inform release decisions. An especially controversial illustration is forecasts used by prison officials to determine which inmates are good risks for unsupervised early release. Other applications include bail recommendations or charging determinations by prosecutors, sentencing decisions by judges, and the nature of supervision provided for individuals on parole.

The pages ahead consider statistical tools to such ends. There have been a number of very recent and important developments well beyond state-of-the-art even five years ago. But context is also considered. The statistical tools will not properly respond to decision-makers' needs if the setting and the consequences of forecasting errors are ignored. Yet in much past practice, they have been.

Criminal justice forecasting has a rich and lively history. In particular, parole decisions in the United States commonly have been informed by forecasts since the 1920s. (*Burgess, 1928; Borden, 1928*). Over the years, varying mixes of clinical judgements and actuarial methods have been used (*Monahan, 1981*). Of late, the greater accuracy and transparency of the actuarial methods (*Hastie and Dawes, 2001: 58-70*) has favored statistical approaches. But often, parole boards are required to use their discretion. As a result, clinical judgments can still play an important role.

Despite their pervasiveness, the accuracy of parole forecasts is difficult to determine. Even less is known about the accuracy of similar forecasts at other criminal justice decision points (*Olin and Duncan, 1949; Glaser, 1955; Dean and Dugan, 1968; Wilkins, 1980; Farrington and Tarling, 1985; Gottfredson and Moriarty, 2006; Berk, 2008a; Skeem and Monahan, 2011*). The most apparent roadblock is that far too few forecasting procedures have been properly evaluated. Many are not evaluated at all despite claims of "instrument validity." Although many are said to be "valid," the methods used and subsequent results are never revealed in meaningful detail, especially when the instrument is "proprietary" (*Berk et al., 2005*).

Even when serious evaluations are reported, the evaluations are often poorly done. It is common, for example, to use the same data to build and test a forecasting procedure. It has been long recognized that this "double-dipping" will make forecasting accuracy appear to be more accurate than it really is (*Ohlin and Duncan, 1949; Reiss, 1951; Ohlin and Lawrence, 1952*). That is just to the beginning. A charitable assessment is that, to date, the accuracy of criminal justice forecasts is mixed at best.

At the same time, the bar is being raised. Recent advances in statistics and computer science are at least in principle setting new standards for forecasting accuracy (*Berk, 2008a*). Aspirations are further bolstered when these tools are combined with the increasing availability of very large datasets containing hundreds of potential predictors. Whatever the past performance of parole forecasts, it may now be possible to do substantially better. It is this promising future that will be emphasized in the pages ahead.

Now may be an unusually good time to capitalize on recent technical developments. There seems to be a growing acceptance of modern actuarial methods for criminal justice decisions that were in the past made by informal assessments of "future dangerousness." Included are bail recommendations and charging, diversion from prison and incarceration sentences, determinations of the security level in which inmates are placed, and the intensity of parole and probation supervision.[1]

---

[1] There are also a number of related applications such using forecasts of wrongdoing to target inspections of industrial polluters, forecasts of child abuse to inform oversight of children in foster

For example, a recent statute in Pennsylvania authorizes the Pennsylvania Commission on sentencing to develop a risk forecasting instrument to help inform sentencing decisions under the state sentencing guidelines. There is an interesting history and many complicated issues (*Hyatt et al., 2011*), but for our purposes the key section reads as follows:

42 Pa.C.S.A.§2154.7. Adoption of risk assessment instrument.

(a) General rule. – The commission shall adopt a sentence risk assessment instrument for the sentencing court to use to help determine the appropriate sentence within the limits established by law for defendants who plead guilty or nolo contendere to, or who were found guilty of, felonies and misdemeanors. The risk assessment instrument may be used as an aide in evaluating the relative risk that an offender will reoffend and be a threat to public safety.

(b) Sentencing guidelines. – The risk assessment instrument may be incorporated into the sentencing guidelines under section 2154 (relating to adoption of guidelines for sentencing).

(c) Pre-sentencing investigation report. – Subject to the provisions of the Pennsylvania Rules of Criminal Procedure, the sentencing court may use the risk assessment instrument to determine whether a more thorough assessment is necessary and to order a pre-sentence investigation report.

(d) Alternative sentencing. – Subject to the eligibility requirements of each program, the risk assessment instrument may be an aide to help determine appropriate candidates for alternative sentencing, including the recidivism risk reduction incentive, State and county intermediate punishment programs and State motivational boot camps.

(e) Definition. – As used in this section, the term risk assessment instrument means an empirically based worksheet which uses factors that are relevant in predicting recidivism.

The content of this legislation would come as no surprise to individuals and organizations following developments in criminal justice forecasts of risk. For example, the Pew Foundation's Public Safety Performance Project recently released a report advocating procedures that "when developed and used correctly ... can help criminal justice officials appropriately classify offenders and target interventions to reduce recidivism, improve public safety, and cut costs" (*Pew, 2011: 1*). A report from the National Center for State Courts (*Casey et al., 2011*) provides more detail. Both reports, however, could have made an even stronger case had they properly summarized current state-of-the-art.

One must not forget, however, that the manner in which such forecasts are received by stakeholders is often part of a highly-charged political process, especially in these times when budget reductions have put enormous pressure on criminal justice agencies to do more with less. For example, a recent OP-ED article in the Los Angeles times, written by the chairman of the state's Senate Labor and Industrial Relations Committee, claimed that forecasts used to determine prison releases were compromising public safety (*Lieu, 2011*).

Beginning in 2011, the California Department of Corrections and Rehabilitation (CDCR) was required under Senate Bill X3 18 to release a "limited number of nonviolent, nonserious and nonsex offenders with no parole conditions and no

---

care, and forecasts of violence to help shape the manner in which fugitives are apprehended by Federal Marshals.

parole supervision." These offenders were selected by CDCR's "computerized risk-assessment model." According to a report from the state's Office of the Inspector General, about 25% of those released were actually "at a high risk for violence." CDCR blames the forecasting errors on missing information on prior record in the forecasting database. But there is no report on how accurate the forecasts were for individuals with complete data.

Forecasting the risk of re-offending raises a number of important and contentious issues. Although this book focuses especially on recent developments in statistics and computer science that promise better forecasts, some legal and ethical matters will be introduced when directly germane.

1.1.1 A Brief Motivating Example

To help motivate the material ahead, consider the following brief example. The data come from a large East Coast city. The relevant population is individuals on probation in that city. The goal of the supervising agency was, in this instance, homicide prevention. Reductions in perpetration *and/or* victimization among those it supervised was the aim.

A forecasting procedure was developed to predict whether an individual under supervision would "fail" while under supervision. The "failure" to be forecasted was whether an individual under supervision was arrested for (1) homicide or (2) attempted homicide, *or* (3) was a victim of a homicide *or* (4) the victim of an attempted homicide. An attempted homicide had the same failure status as a completed homicide. All other probation outcomes were treated as "successes." The follow-up period was 18 months.

Predictors were drawn from the usual sorts of administrative records to which such agencies have access. Among the predictors were age, prior record, gender, the most recent conviction offense, and the age at which an individual first appeared in adult court. There are certainly no surprises. In addition, all possible interaction effects were included up to the equivalent of about 6-way product variables. These can identify unanticipated structure in the data that would be missed by conventional approaches and can dramatically improve forecasting accuracy. In effect, there was well over 300 predictors, most of which were not identified in advance, and if they were, would have had no clear substantive interpretations.[2]

Because the outcome to be forecasted was binary, a machine learning classifier was used to construct the forecasting algorithm. The relative costs of forecasting errors build into the algorithm were asymmetric. A failure to correctly identify a high-risk probationer (i.e., a false negative) was viewed as far more costly than incorrectly identifying a probationer as high risk (i.e. a false positive), and the cost ratio of false negatives to false positives was set at 20 to 1. False negatives were

---

[2] This is a lot like much recent work in bioinformatics where there is a lot of data but not that much theory to go with it (*Baldi and Brunak, 2001*). Microarrays and gene expression is a good example.

treated as 20 times more costly. The agency was, in this instance, very concerned about homicides that could have been prevented.

	Forecasted Not Fail	Forecasted Fail	Accuracy
Not Fail	9972	987	.91
Fail	45	153	.77

Table 1.1 Confusion Table for Forecasts of Perpetrators and Victims. True negatives are identified with 91% accuracy. True positives are identified with 77% accuracy. (Fail = perpetrator or victim. No Fail = not a perpetrator or victim.)

For these data, about 2% fail within 18 months. From a statistical perspective, the base rate is very low. It was challenging, therefore, to gain much forecasting leverage from the predictors. If one always forecasted that there would be no failures, the forecast would be correct nearly 98% of the time. It is difficult to imagine a forecasting procedure using predictors that could do better.

But forecasting solely from the base rate would mean that all true positives would automatically be false negatives. Because of the very high costs of false negatives relative to false positives, this was an unsatisfactory approach. It was important that predictors be used to differentiate between those more likely to fail and those less likely to fail. In other words, overall accuracy could be somewhat compromised if in trade a substantial number of high risk individuals were correctly identified as such.

Table 1.1 shows some results in a "Confusion Table" assembled as the forecasting procedure was being developed. The table is nothing more than a cross-tabulation of actual outcomes and forecasted outcomes. The forecasts are real in the sense that the table was built from data not used as the forecasting procedure was formulated. That is, a proper confusion table is assembled from a "hold-out" sample containing predictors and the outcome to be forecasted. The table is an honest rendering of forecasting accuracy.

There are 198 "failures" overall. The forecasting algorithm correctly identified 153 of the 198 failures, for an accuracy rate of 77%. There are 10,959 "successes" overall. The forecasting algorithm correctly identified 9972 of the 10,959 for an accuracy rate of 91%. These figures represent how well the algorithm was able to correctly identify which offenders failed and which did not using information contained in the predictors. By exploiting that information, the algorithm ideally can find a group composed largely of individuals who succeeded and another group composed largely of individuals who failed. When this happens, the procedure has promise as a forecasting tool.

Given the high rate of accuracy despite the very low base rate, decision-makers felt that the forecasting algorithm had potential. Refinements to the algorithm improved its performance, and it was later installed on an agency server to inform supervisory decisions. That software is now routinely used to inform probation supervisory decisions.

There is certainly a lot more going on in Table 1.1. For example, how many false positives are there for every true positive — how many individuals are incorrectly identified as high risk for every individual correctly identified? For the moment, there are two general points to be made.

First, accurate forecasts are often very easy to construct. When one of a pair of outcomes is rare, forecasting the more common outcome can lead to very accurate predictions. Usually, however, we seek more nuanced results that respond to the costs of forecasting errors. With large databases and modern forecasting machinery, one can still do quite well.

Second, although statistical concerns and policy concerns can be conceptually distinct, they should inform one another in practice. Indeed, there is really no such thing as a "pure" forecast. Even when researchers think they are proceeding in a value-free manner, they are implicitly making value-influenced choices that can have important effects on the forecasts. These choices cannot be avoided, and need to be explicitly aired.

Chapter 2
Some Important Background Material

Abstract Statistical forecasts of criminal behavior are far more than a technical challenge. They inform real decisions by criminal justice officials and other stakeholders. As a result, a wide range of issues can arise. Although for expositional purposes they must be examined one at a time, in practice each should be considered as part of a whole; decisions about one necessarily affect decisions about another. We begin with some rather broad concerns and gradually narrow the discussion.

2.1 Policy Considerations

Criminal justice forecasts are meant to inform real decisions. As such, they are embedded in a variety of political and administrative considerations. These considerations are a useful place to start because they are so central to the forecasting enterprise.

2.1.1 Criminal Justice Forecasting Goals

Forecasts of re-offending are used in settings where several masters must be served. The most salient and widely-stated consideration is public safety. Individuals who pose a significant risk to the public must be accurately identified so that their future crimes can be prevented. Prevention failures, especially when the crimes are unusually heinous, are tragedies for victims and their families, but also have significant costs for the criminal justice decision-makers responsible. One important result is that these decision-makers are often especially sensitive to risk when it comes to crimes of violence. This has important implications for how forecasts are made and used, as we will see later.

In recent years, a second concern has become increasingly important: resource allocation. Forecasts can be used to allocate criminal justice resources more efficiently. For example, probation supervision can vary in content and intensity. Cost varies accordingly. Ideally, forecasts of future dangerousness can be used to allocate scarce resources to those most in need (*Berk et al., 2009*). Similar issues can arise in the allocation of prison beds at different security levels (*Berk et al., 2006*). Beds in high security facilities are very costly and should only be assigned to inmates who pose a significant threat to themselves, other inmates, and/or prison staff.

A third matter is transparency. Decision-makers should be able to understand how forecasts are constructed, as well as their strengths and weaknesses. Stakeholders should be able to understand how the forecast-informed decisions are made. One of the advantages of actuarial methods is that, in principle, they are replicable and therefore transparent. However, transparency is not only a matter of the actuarial methods employed. It is also a matter of what is disclosed and the form in which the information is conveyed. Providing overwhelming amounts of information without adequate structure or documentation is not transparency.

Linked to the goal of transparency is the goal of fairness. The issues are complicated. For example, if race is a powerful predictor of violent crime, and if perpetrators tend to commit those crimes against people like themselves, should race be used as a predictor (*Berk, 2009*)? What about possible surrogates for race such as neighborhood characteristics? And if race is suspect, what about age and gender? Determining in this context what is fair can require balancing of difficult trade-offs. For example, young African-American men disproportionately kill other young African-American men. By not including age, gender, and race as forecasting predictors, one may be sparing some young African-American men substantial time in prison, but at the cost of the deaths of other young African-American men.

Then, there is the critical matter of practicality. Criminal justice forecasting procedures that cannot be implemented in a timely and effective manner serve no policy purpose. Stated a bit differently, if forecasts aim to assist criminal justice decision-makers, they must be sufficiently accurate and simple to obtain in real time. Fortunately, actuarial methods can be easily implemented on commonly-available computers. The primary hurdle is getting the requisite data to those computers in a timely fashion. But, that rarely poses a significant problem because many criminal justice agencies have IT staff that can readily assemble the data when needed.

Finally, some forecasting instruments used in criminal justice settings make "needs assessment" the primary goal. This commonly means that the deficits of individuals in custody are to be identified so that appropriate interventions can be introduced. The process is less about forecasting and more about treatment. A key implication is that the instrument used is supposed to be able to determine the *causes* of an an offender's anti-social behavior. Behind the scene is typically a causal model of offending. Sometimes the causal model is statistical and based on causal modeling traditions in the social sciences. Other times the causal model is intuitive, or based on social science theory and/or craft lore.

There is nothing wrong in principle with combining forecasts of offending with determinations of offender needs. But in practice, it is usually better to separate the two activities. Then, statistical tools can be optimized to best serve their stated purposes. For instance, forecasts can be tuned to maximize forecasting accuracy using at least some variables that have no causal role. To take an example that is not as silly as it first sounds, if an offender's shoe size is a useful predictor, it can be productively included. Yet, it is hard to imagine what a causal connection to re-offending would be, let alone what an appropriate intervention would entail. In this book, therefore, we concentrate on forecasting tuned to maximize forecasting accuracy alone. Any insights about offender needs are a bonus. In contrast, causal models of offending used in assessments of need should make causal explanation primary. Then, meaningful forecasting accuracy is a bonus.

2.1.2 Decisions to be Informed by the Forecasts

There is a range of decisions that criminal justice forecasts can inform. Among the decisions are:

- bail recommendations;
- charging by prosecutors;
- sentencing, including probation;
- placing an individual in an appropriate prison security level;
- whether to release an individual on parole; and
- the kinds of supervision and services to be provided for individuals on probation or parole.

It is likely that with different decisions, there will be different datasets available. This implies the need for different forecasting procedures and forecasts tuned to different priorities. Unfortunately, one finds too often that very different criminal justice decisions are based on the same forecasting instrument. For example, a given kind of threat assessment instrument may be used in sentencing and in parole decisions despite the obvious fact that the populations, legal and administrative requirements, available data, and outcomes to be forecasted can be dramatically different. There is no reason to believe that even if a threat assessment instrument works well for sentencing, for instance, it will work well for parole decisions. It is also possible that forecasts will be suboptimal for both because they were tuned to neither.

In addition to the kind of decision, *when* in the process the decision is made matters. For example, one may want to arrange parole supervision regimes differently depending on the information one has at intake compared to the information one has a year later. Presumably, the amount and quality of information improves as time goes by. The options available for interventions can change as well. Perhaps a drug treatment program that was fully subscribed at intake has openings six months later. The point is that, once again, one size does not fit all. The forecasting procedures used should be designed specifically for the kind of decision and its timing.

2.1.3 Outcomes To be Forecast

Although criminal behavior is usually the outcome that criminal justice officials want to forecast, the behavior in question can be defined in many ways. Historically, the most common outcome is whether or not there was any arrest for a new crime and/or a significant violation of parole or probation rules. Of late, there is increasing interest in forecasting different kinds of crime and/or violations, such as crimes of violence or sex crimes. But that necessarily raises the question of "compared to what?" For example, if committing a murder is defined as the "failure" of concern, a "success" might range from a brutal armed robbery or aggravated assault, to no crimes whatsoever.

Among the outcome possibilities to forecast are:

- An arrest for murder or attempted murder compared to all other possible outcomes — two outcome categories;
- An arrest for a violent crime compared to all other possible outcomes — two outcome categories;
- An arrest for a violent crime, versus an arrest for a crime that is not violent, versus no arrests at all — three outcome categories;
- All FBI Part I arrests, versus all FBI Part II arrests, versus no arrests at all — three outcome categories;
- An arrest for sex crimes compared to all other outcomes — two outcome categories;
- An arrest for perpetration of a homicide or an attempted homicide as well as victimization by a homicide or an attempted homicide compared to everything else — two outcome categories; and
- Person crimes, versus property crimes, versus "victimless" crimes, versus no crimes — four outcome categories.

A range of other options can be defined if "arrest" is replaced by "conviction," and if violations of parole or probation conditions, or a failure to appear in court, are relevant. Usually, the choice of what to forecast is a blend of legal, political, and technical concerns. For example, we have already mentioned the low base rate problem that follows when the outcome to be forecasted is rare. But a relatively infrequent crime, if seen as a major threat to public safety, may be the major forecasting target. Another common issue is resource constraints. If the outcome to be forecasted is not especially rare and an allocation of significant resources for that forecasted outcome is planned, an outcome definition that includes too many cases can outstrip agency capacity.

2.2 Data Considerations

Forecasts of re-offending depend on having appropriate data. It is useless to advocate for greater use of criminal justice forecasting unless such data are available. But what qualifies as "appropriate" is multidimensional and complicated.

Criminal justice forecasts always have at least implicitly a target population for which forecasts are to be made. For example, the target population may be individuals in a particular jurisdiction over a particular period of time for whom bail recommendations are required. Data from that population should be used in developing the forecasting procedures and to evaluate how those procedures perform. And then, individuals for whom forecasts are desired should be from that population as well. So much for textbook pronouncements. Real applications can get messy in a hurry.

2.2.1 Stability Over Time

All forecasting requires that the future to be forecast is like the past from which the forecasting procedure is built. Broadly understood, the point is obvious. But the future is never exactly like the past, and change can come in many forms, at different speeds, and with different consequences for forecasting accuracy. As in so many things, the devil is in the details. We need to go there — at least a bit.

Decision-makers must operate within a formal structure of statutes and administrative regulations. For example, sentencing guidelines can be introduced, revised, or abolished. Certain kinds of crimes, such as sexual offenses, can be singled out for special treatment. The formal structure can also affect how re-offending occurs and how it is defined. Certain crimes can be changed from misdemeanors to felonies. The implications of a failed drug test can be altered. All such changes in the formal structure make the future different from the past. The question, therefore, is how much forecasting accuracy is affected, not whether it is affected at all.

The identities of decision-makers also matter a great deal. Parole board personnel turn over. The mix of probation or parole officers on the street is continually changing. Sitting judges lose elections to newcomers or retire. All such processes can affect forecasting accuracy. Again, it is a matter of degree.

Factors that can encourage or discourage crime also vary. The job market is thought to be important. So are schools. The same applies to rivalries or turf battles between gangs. There are also changing crime opportunities. Markets for different kinds of drugs is a common illustration. The resources and effectiveness of criminal justice agencies matter as well. Examples include the number and deployment of police officers and the size of the incarcerated population. There is hope that various kinds of diversion and rehabilitation programs will help, although demonstrable successes can be hard to find.

Finally, there is the changing demographics of law enforcement jurisdictions. Although young males from low income households do not have a monopoly on violent crime, they surely dominate the statistics. It follows that when there are more of them in an area, the risk of serious crime increases. Among the many processes that can affect these numbers is the kinds of households that are moving in and out. For instance, gentrification can reduce crime in a jurisdiction.

There is no way that changes in all of the factors just mentioned can be fully tracked; their impacts on forecasting accuracy are difficult to anticipate in any case. However, major changes in the statutory or administrative setting are easy to identify and will often mean that forecasting procedures tuned before the changes will need to be revised. There may also be important public relations and political reasons to update the procedures. It is easy to imagine the difficulties if a forecasting procedure built under one set of sentencing guidelines were being used to make forecasts under another. The forecasts and underlying procedures become almost an irresistible target.

For the other factors, it may be best to just track the accuracy of forecasts over time. If accuracy is degrading, there is evidence that the values of one or more factors are drifting or that the causal processes have evolved. When the reductions in accuracy become large enough to matter, it is time for a revision. What little experience there is with the kinds of forecasting discussed later suggests that the forecasting procedures usually degrade slowly and not enough to matter for 3 to 5 years, perhaps longer. But there can be important exceptions when the formal structure of decision making (e.g., key statutes) are substantially altered.

In short, although the future will never be exactly like the past, a major consideration when forecasting procedures are constructed is whether forecasting accuracy is acceptable. This will usually mean that in a head-to-head contest of forecasting accuracy, the new methods perform better than the old. Less clear is *when* to revise the current forecasting methods. Important changes in statutory or administrative structure will usually require an update. In the end, however, substantial reductions in forecasting accuracy are probably the most direct indicator.

Finally, there is no universal way to determine when reductions in forecasting accuracy are "substantial." It depends on the consequences of forecasting errors and how many there are. Is failure to predict, say, 10 homicides worse than failure to predict 100 burglaries? What about 10 homicides compared to 1000 burglaries? These kinds of relative valuations should be derived from the preferences of stakeholders whose views can vary widely. We will consider such issues in depth later, but suffice it to say that the relative costs of different kinds of forecasting errors are critical to the proper construction of forecasts, to how those forecasts are used, and to when they need to be revised.

2.2.2 Training Data and Test Data

An essential component of building a good forecasting procedure is to work with two datasets: (1) a random sample of training data from the relevant population, with which to develop the procedure; and (2) a random sample of test data from the same population, with which to evaluate the procedure. Both datasets should include the same predictor variables and response variables whose values one ultimately wants to forecast.

One of the frustrating errors in past criminal justice forecasting was to rely on procedures constructed and "validated" with the same data. Most forecasting methods are tuned to the data on which they are developed. Many formally maximize some measure of fit between what the forecasting method predicts and what the data show. Logistic regression is a ubiquitous example. Model construction enhancements such as stepwise logistic regression can make the tuning more effective. However, tuning responds not just to systematic features of the data that can help forecasting accuracy, but to idiosyncratic features of the data having no predictive information. One result is that the model "overfits," and stakeholders can be victimized by unrealistic expectations. When called upon to forecast from new data, accuracy may be very disappointing..

It is important, therefore, to have a second dataset. This should ideally be a random sample from the same population, with which true forecasts can be made. That is, the forecasting methods developed with the training data are evaluated with the test data. For example, a forecasting algorithm constructed with the training data is moved to the test data. There, the values of the predictors are inserted to construct forecasts of the response variable. Like the training data, the test data contain the observed outcome for each case. This allows the forecasted values to be compared to the observed values of the response variable. Then, an honest evaluation of forecasting accuracy can be undertaken in conventional ways. Commonly, some function of the forecasting errors is used.

2.2.3 Representative Data

Having training and test data that are random samples from the population for which forecasts are to be made is not by itself sufficient. One must consider carefully the nature of that population. For example, training and test data from a population of individuals released on parole is not likely to be appropriate for forecasts informing bail release recommendations. Parolees have typically served substantial time behind bars. Bail recommendations are made well before guilt for a felony crime has been officially determined. Parolees, therefore, will likely differ in their exposure to incarceration and whatever impact prison programs might have. They are on the average likely to be older and have longer prior records, as well as convictions for more serious crimes. A forecasting procedure tuned to parolees may provide seriously misleading forecasts if applied to bail recommendations. Similar problems can arise if a forecasting procedure tuned to individuals released on bail is used to inform sentencing decisions. The relevant populations differ.

One must distinguish, however, between the values of predictors that can on average vary over populations, and the manner in which those predictors forecast. To take a simple example, age may have much the same impact in a sample of parolees and a sample of probationers. That is, the relationship between age and reoffending is effectively the same. In both populations, older individuals are less likely to reoffend, other things equal. One might find that, perhaps, for each 5 years of age, the probability of reoffending declines by 4%. But because parolees tend to be older, they will on average have a lower risk for reoffending, other things equal. In short, the forecasting procedure is performing as it should.

In contrast, it may be that for parolees, the probability of reoffending declines by 7% with every 5 years of age, whereas for probationers the reoffending declines by 2% with every 5 years of age. Then, the same forecasting procedure should not be used for both populations because the algorithms are not the same. If in the data one can determine which individuals are parolees and which are probationers, there is no problem, at least in principle, to construct forecasting procedures for each separately.[1]

There is also the matter of unobserved outcomes for the relevant population. Suppose a forecasting procedure is being designed to inform bail release decisions. A population of individuals with earlier bail decisions is found for which there are predictors from routine administrative records and rap sheets with which to measure new crimes before trial. One could draw two random samples from this population: one for use as training data and one for use as test data. With these data in hand, it would be possible to examine factors associated with new arrests of various kinds for those individuals released. The goal would be to determine what sorts of individuals should not have been released.

However, there is a complementary goal: what sorts of individuals were *not* released but would have been good risks? There is no way with these data to directly address that question because for individuals not released, there was no opportunity to see how well they might have done. The outcome data are "censored" for individuals who remain in jail. Censored outcomes for particular classes of individuals are a general problem in criminal justice forecasting when there is, for whatever reason, no way to observe an outcome of interest. The usual "solution" is to try to find data that are not censored for individuals like those whose data are censored. But in practice, this is very difficult to do. Often, there are no data that can pass a sniff test. In the rare instances when promising data are found, there will be observed differences of unknown importance (e.g., military service) and unobserved differences about which one can only speculate.

2.2.4 Large Samples

Substantive understanding can have different empirical foundations from forecasting. When the intent of an empirical study is to explain how predictors are related to a response, samples of modest size will often suffice. Interest typically centers on the relatively few important predictors of particular substantive interest. Predic-

---

[1] Additional complications can arise if the response function is nonlinear. For example, if the forecasting procedures is build using data on parolees, there may be for young offenders seeking bail release an insufficient number young parolees to determine how age is related to reoffending.

tors having small and perhaps barely detectable relations with the response can be discarded with little penalty in understanding. As long as parameter estimates are obtained with sufficient precision, there is no need for a large sample. A few hundred observations each in the training data and test data may well suffice.

When the intent of an empirical study is to forecast, the bigger the sample the better, at least typically. There will often be a large number of predictors, each having a very weak relationship to the response. But, *as a group*, they can noticeably improve forecasting accuracy. There may be little or no substantive story in such predictors — predictor by predictor they hardly matter. Yet, they are systematically associated with the outcome, and are numerous enough to make a difference in the aggregate. Under such circumstances, it can make good sense to sever any connection between explanatory methods and forecasting methods. If the goal is accurate forecasts, data collection and data analysis should be guided by that purpose.

The ways to capitalize on the information in very large samples will be considered in some detail later within the favored machine learning methods introduced. Suffice it to say that one must have hardware and software that are not overwhelmed, and a combination of predictors and statistical procedures that can find even small pockets of systematic structure. Samples of well over 100,000 can then be productively exploited.

2.2.5 Data Continuity

The point of having training data and test data is to build the forecasting procedure. However, use of the forecasting procedure depends on the data available for new cases that are not part of the procedure building process. It stands to reason that for these cases, data in the same form must continue to be available. The presence of new predictors is not a problem. They can simply be ignored. Problems develop if earlier predictors are no longer available or are measured in new ways. Changes in data format or documentation can lead to at least temporary difficulties.

It is unlikely that the data available will be completely stable for long periods of time. Therefore, documentation of the data must be revised as the data change so that where possible, misunderstandings are avoided. For example, for the racial category "white," hispanics may at some point be represented in the data as a separate ethnic/racial category. The meaning of "white" is no longer the same. Eventually, there will likely be so many changes that is necessary to reconstruct the forecasting procedure using the most current data available. This is not necessarily bad. The newer data may have more potential predictors and better measured predictors.

Much as for the complications introduced by lack of stability in the relevant populations, there will usually be no definitive way to determine when to revise the forecasting procedures. Occasionally, a few key predictors will no longer be available or are measured in new ways. Employment history is a common example. But more often, degradation of the predictors with which a forecasting procedure was built will cause a decline in forecasting accuracy. At some point, forecasting accuracy will become unacceptable.

2.2.6 Missing Data

Sometimes variables used to develop forecasting procedures have missing data. When for a given variable all of the data are missing, there is nothing to be done unless the missing data can be obtained. There is no statistical fix. For example, if the field for prior arrests is totally empty, the choice is between not including that variable in the analysis or obtaining that information from other sources.

Very commonly, however, missing data are scattered around in the data. There are fields for some cases and some variables that have no information other than a missing data code. Missing data may be represented by a "blank" or some affirmative code such as NA or -99. Sometimes, there is a bit more information such as whether a missing value results from a justified exclusion, an explicit regulation that precludes obtaining the information, an oversight, or some other cause. A missing data code of -99 might mean one thing, and a missing data code of -98 might mean something else.

An example of a justified exclusion would be measures of performance in a prison job training program when the individual had no opportunity to enroll. However, if one knew more about the reasons why there was no opportunity to enroll, there could be some predictive information (e.g., the inmate was often insubordinate). An example of an explicit regulation that precludes obtaining certain information would the sealing of juvenile records. An oversight might mean that the requisite information was never obtained or that there was some subsequent error in how the information was recorded and processed.

When the missing data are scattered around, there are at least five possible responses that can vary by practicality and consequences. First, one can revisit the primary documents from which the data were constructed to see if the missing values can be determined. Sometimes, missing data results from an honest mistake or oversight easily corrected from source materials. This solution is ideal.

Second, one can discard entire cases when for those cases there are any missing values for any variables. This is sometimes called "listwise" deletion. If such cases are few, nothing of a real importance is likely to be lost. If such cases are many *and* such cases can differ systematically from the cases with complete data, a lot can be lost. For example, inmates serving shorter sentences may be less likely to have full "work-ups" or opportunities to participate in prison programs. If these case cases are dropped, forecasts are being constructed for offenders convicted of more serious crimes. At the very least, such limitations must be explained to stakeholders. There could be implications for the generalizability of the forecasts.

Third, variables (not cases) with missing data can be dropped. Continuing with the inmate example, predictors associated with prison jobs can be discarded when the time comes to construct the forecasting procedures. If such variables have little predictive value, there is effectively no cost. If such variables have substantial predictive power, one must decide whether the forecasts are sufficiently accurate nevertheless. It may also be possible to revisit the data and find predictors without missing values that can serve as useful surrogates. For example, employment history prior to incarceration may be a good surrogate for performance in prison job training programs.

Fourth, there are a number of imputation procedures that can be employed. These are beyond the scope of this book and for machine learning approaches may well be unnecessary. Imputation is usually undertaken by extracting information from variables for which missing data is not a problem. To take an obvious example, a conviction requires that there was an arrest. Consequently, one can infer that if there is a prior conviction, there is prior arrest even if no arrest is recorded. But if prior convictions are included as predictors, one already is using the information convictions contains. In machine learning, moreover, one can include a very large number of predictors in a fashion that allows for high order interactions and highly nonlinear relationships with the response. There will likely be little information not already exploited for forecasting that could have been used for imputation.

Finally, sometimes there are ways to code variables so that missing data problems are averted. For example, there may be for many observations no information on juvenile arrests. A reasonable inference for such cases may be that there were no juvenile arrests. A count of such arrests would be coded as zero. Predictors that are factors provide other kinds of useful options. In particular, one can define a new factor category for "missing" and analyze the factors just as one ordinarily would. If the missing data has a systematic explanation, the missing data category can contain predictive information. For example, military service could have four categories: currently serving, served in the past, never served, and missing. Perhaps those with no information on military service are low (or high) risk? "Missing" then becomes a legitimate category for military service.

In summary, the best situation is to have very little missing data to begin with. Experience with the sorts of records used to forecast future dangerousness suggests that this is often the case. Because such records are used for administrative purposes, there can be a concerted effort to make those records accurate and complete. But if there are substantial amounts of missing data that cannot be obtained from source documents, probably the best strategy whenever possible is to recode the problem away. A reasonable second choice is to drop the offending predictors. It is important to keep in mind that forecasting, not explanation, is the primary goal.

2.3 Statistical Considerations

It is important to start with a broad view of the statistical tools used to generate forecasts. Although we will focus later on machine learning, one can think of machine learning in a forecasting context as a special case of actuarial methods. We have already mentioned actuarial methods in passing. It is now time to consider actuarial methods in somewhat more depth.

2.3.1 Actuarial Methods

Actuarial methods are statistical and mathematical procedures commonly used in public health, banking, insurance, and finance to assess risk. By "risk" one means the prospect of some undesirable outcome and as such, combines the chances of that undesirable outcome with its unwelcome consequences. An event certain to occur but with no untoward consequences has no risk. An event that cannot occur does not entail risk no matter how dire the possible consequences. In this book, actuarial methods are the focus.

Much like such insurance companies, banks or investment firms, criminal justice agencies are in part risk management enterprises. Desirable and undesirable outcomes must be effectively anticipated. Broadly speaking, determining the chances that an individual with a particular profile will commit a serious crime is not very different from determining the chances that an individual with a particular profile will be involved in a fatal automobile accident, contract a chronic disease at an early age, or will own property damaged by a hurricane. Likewise, it is not very different from determining the expected returns from a loan to a small business, a mortgage to a new home buyer, or an investment in an initial public offering. It is all about anticipating the future with a useful level of accuracy. There must be sufficient forecasting "skill."

Insurance companies, for instance, make such forecasts to help determine what premiums to charge customers. Different kinds of customers come with different risk probabilities, and these risks come with different amounts of loss. Risks with higher probabilities and/or associated with more costly losses can justify higher premiums. Consequently, insurance companies must integrate their forecasts with the costs of particular kinds of events. Forecasts alone, even very accurate forecasts, will not suffice. The same lesson applies to criminal justice applications. Even very accurate forecasts are by themselves insufficient. The consequences of the behavior being forecasted must be introduced if risk is to be properly addressed.

The list of relevant statistical and mathematical tools used as actuarial methods is long. It can include simple cross-tabulations to complex models and modern data mining. In most settings, for example, a cross-tabulation of gender by whether a parolee is arrested for a crime of violence within two years of release from prison will show that men are substantially more likely to fail. This can be useful to know and very easy to ascertain. At the other extreme of complexity can be data mining procedures that link various parole outcomes to complex parolee profiles by searching for patterns in a very sophisticated manner over hundreds of thousands of cases.

Actuarial methods can also differ in the degree to which they try to represent mechanisms driving the outcome. The relevant causal models popular in the social sciences are typically statistical theories of criminal behavior. If a causal model is a good one, it will forecast well — or at least that is the common assumption. But one can productively separate forecasting an outcome from explaining why that outcome occurs.

Because data mining leads to forecasts that do not depend on causal explanations of the outcome, some use the term "algorithmic" for the data mining approach to forecasting. A forecast is generated by statistical machinery whose sole purpose is to maximize some function of forecasting accuracy. There is no model in the usual social science sense and no necessary account of why the criminal behavior did or did not occur. In the pages ahead, machine learning, a particular kind of data mining, will be the favored approach.

The outcomes to be forecasted will be categorical. They usually will be some form of criminal activity or a violation of rules imposed as a condition of the intervention applied (e.g., failing to report to a probation officer). For categorical outcomes, actuarial methods are sometimes called "classifiers." Their goal is to assign cases to classes. For example, should a parolee who is a 25-year-old male with 3 prior felony convictions and an affiliation with a street gang be assigned to the class of individuals likely to commit a violent crime while under supervision? Machine learning methods used with categorical outcomes are almost always called classifiers.

Whatever the actuarial methods preferred, it is at present hard to find what one needs off-the-shelf. Or put more positively, the forecasting results are likely to be better if the off-the-shelf methods are a starting point for more refined procedures hand-tailored to the circumstances in which the methods will actually be used. In this book, therefore, we will talk about forecasting approaches that must be developed on a case-by-case basis, honestly tested in the settings where they will be used, implemented so that they can be employed when needed, and integrated into actual criminal justice decisions. Actuarial methods are an essential part of the mix, but hardly all there is.

2.3.2 Building in the Relative Costs of Forecasting Errors

Decision-makers readily understand that different kinds of forecasting errors can have different consequences and that their costs can vary enormously. A forecast that fails to identify a prospective murderer is likely to be far more costly than a forecast that fails to identify a prospective burglar. Such information should be an important decision-making factor. Likewise, decision-makers readily understand that some behavioral outcomes are far more likely that others. This information should be an important decision-making factor as well. Both kinds of information are used in proper assessments of risk.

Thus, risk has two components: the probability of a given outcome and the expected costs of that outcome. Consider first the costs. The chances of a new crime may be the same for two individuals, but for one that new crime could be a crime of violence whereas for the other that new crime could be a property crime with no violence. When, as usual, the former is seen as more costly, the individual projected to be violent will likely experience an intervention that is has stronger incapacitative content. Broadly speaking this can make good sense. The costs of the crime are higher.

Now consider the chances of a new crime. The costs of a given crime may be the same for two individuals, but one may be much more likely to commit that new crime. That individual will likely experience an intervention with stronger incapacitative content. Broadly speaking, this too can make good sense. The chances that the crime will be committed are greater.

As a formal matter, the risk associated with a given outcome is defined as the product of the probability of that outcome times its expected costs. For criminal justice decision making, it is important that the probability be estimated as accurately as possible. This goes directly to the statistical material included in later chapters.

In practice, decision-making can often proceed with only *relative* costs being estimated. For example, one outcome is five times more costly than another outcome. Moreover, one can often be satisfied with a decision-maker's beliefs about the relative costs when better information is not available. For instance, how would one put a dollar value on the political consequences of releasing an individual on parole who then commits an especially heinous crime? How would one put a dollar value on the psychic costs for the crime victim's family?

If risk is to be an important factor in decision-making, the next issue is when that information should be used. Usually, it is introduced far too late in the decision-making process. A critical point follows: *the costs of forecasting errors need to be introduced at the very beginning when the forecasting procedures are being developed*. Then, those costs can be built into the forecasts themselves. The actual *forecasts* need to change in response to relative costs.

Recall that conventional statistical methods in principle take a useful step toward this goal. For example, a statistical model for forecasting may be the product of a least squares or a maximum likelihood estimator. There is a loss function to be minimized. The model builds a form of cost minimization. So, what's the problem?

One problem is that costs are not explicitly linked to forecasting errors, although under certain assumptions they can be. A more serious problem is that loss functions for all of the popular statistical methods applied to criminal justice forecasting use *symmetric* loss although for most most decision-makers and stakeholders, the costs are asymmetric.

Consider a forecast of murder. One error is failing to identify a prospective murderer. The term "false negative" is often applied. Another error is to falsely identify an individual as a prospective murderer. The term "false positive" is often applied. For a false negative, there is a homicide that might well have been prevented along with obvious costs to the victim and the victim's family. The credibility of the decision-makers and their organizations can also be compromised. For a false positive, a long prison term might be imposed unnecessarily. There are obvious costs to the individual and to the individual's family. There are also the thousands of wasted dollars spent on incarceration. Clearly, both kinds of forecasting errors have costs. But it is unlikely that the costs will be identical, or even similar.

Some false negatives and false positives are virtually inevitable in practice, and it seems that for criminal justice decision-makers and the public, the cost of homicides that might have been averted is especially dear. It follows that the greater the costs of false negatives relative to false positives, the weaker the forecast of murder can be. That is, more uncertainty can be tolerated. Any hint in a forecast that an individual will commit a violent crime can be sufficient evidence for a lengthy incarceration. It can also justify very intrusive and intense supervision on probation or parole. And it can justify a recommendation to deny bail.

The reverse is also possible, even if unlikely. If the costs of a false positive are far greater than the costs of a false negative, one would want a very convincing forecast of murder before applying strong counter-measures. One would want to be almost certain.

The reasoning just introduced implies that the loss function associated with any forecasting procedures should usually be *asymmetric*. And if this is done, the forecasts made should take that asymmetry into account. The forecasts will (and should) *differ* depending on how the relative costs of false negatives and false positives are handled. There are several principled ways in which this can be done, but some are better than others. We will later return to these issues in a far more formal manner and in considerable depth.

There can also be important constraints on the overall forecasting results that should be introduced when the forecasting procedure is being developed. For example, there may be some politically-acceptable upper bound to the fraction of individuals predicted to commit violent crimes after release on parole. A procedure that identifies, say, 50% of all individuals considered for parole as a significant threat to public safety may strike some important stakeholders as absurd on its face or a symptom of a broad failure of incarceration to either deter or rehabilitate (whether or not that is true).

There can also be resource constraints. An agency in charge of probation supervision may wish to place all individuals forecasted to be dangerous under probation officers with much smaller case loads and with access to a rich mix of support services. But more intensive supervision is more costly, so there is likely to be some limit on the number of such probationers the agency can handle. Forecasts need to be altered to capture this reality. Such constraints needs to be taken into account when the forecasting procedure is built, not at the end when supervisory decisions are made.

The sorts of political and resource constraints just mentioned can be formulated as factors that affect the costs of forecasting errors. For example, if too many false positives imply too many forecasted failures, there can be serious blowback from key stakeholders. There is, therefore, some threshold in the number of false positives above which the costs of false positives sharply increase.

The basic point is this: when, as is typically the case that the costs of false negatives and false positives are not the same, the entire forecasting approach needs to take asymmetric costs into account. The alternative — introducing asymmetric costs after a forecast is made to affect how that forecast is *used* — risks unprincipled and suboptimal decisions. It also means that most of the supporting output from the forecasting procedure is tuned to the wrong policy preferences.

2.3.3 Effective Forecasting Algorithms

The requirements and constraints just discussed have important implications for the kinds of actuarial methods employed. The method chosen must have forecasting accuracy that is among the best, and forecasting accuracy can vary dramatically across different procedures. But, for real-world applications, there are other important criteria. We turn to those now.

There must be a principled way to introduce asymmetric costs that can help shape the forecasts and all other output of interest. Some procedures do not permit the introduction of asymmetric costs. Some allow asymmetric costs to affect the forecasts but not other output of interest. A few allow asymmetric costs to influence the forecasts and all other output.

The procedure must produce replicable results. For a given dataset, the forecasts generated by one analyst can be reproduced by another. Because most actuarial methods depend in part on judgment calls, "replication" means that once the judgment calls of one analyst are conveyed to another, the results are effectively the same.

The procedure must be able to properly take advantage of the kinds of training and test data available. These days, effective forecasting procedures need to be able to exploit very large datasets with tens of thousands of cases (or more) and hundreds of potential predictors. For much of the statistical work in the social sciences, there can be diminishing returns from large data bases. A sample of several thousand cases may well be sufficient when the goal is causal modeling. Explanation and understanding usually depend on finding dominant patterns in the data. Forecasting is different, especially when the outcome is statistically rare. A very large number of observations can really help. If the problem is to find a needle in a haystack, you first need a haystack.

Beyond the forecasts themselves, supporting output is useful to help diagnose problems in the forecasts produced, suggest how the forecasts could be improved, and justify the forecasts to stakeholders. For example, it is important to learn which predictors are driving the forecasts, not because a causal account must be provided, but because strongly counterintuitive results can signal problems with the forecasting algorithm and undermine the use of forecasts that are otherwise quite accurate. To take a simple example, age should be related to recidivism within the usual population of parolees. Older individuals usually pose fewer risks. If age is not related to risk, it could be important to revisit the training data and the actuarial methods applied.[2]

---

[2] Some might argue that these kinds of claims provide a role for social science theory. However, that depends on what one means by theory. With respect to the role of age, how many geriatric offenders are found in crime movies since the 1930s? In popular culture at least, serious crime is

Figure 2.1 is an example of useful output showing the forecasting importance of predictors, which for present purposes can be described very briefly. The training and test data are samples of individuals released from prison on parole. The response variable is three categories of performance on parole measured over 2 years: being arrested for a violent crime, being arrested for a crime but not a violent one, and not being arrested at all. Most of the predictors are the sort commonly used and for now do not need to be explained except that "Charge Record Count," "Recent Report Count," and "Recent Category 1 Count" refer to misconduct in prison.

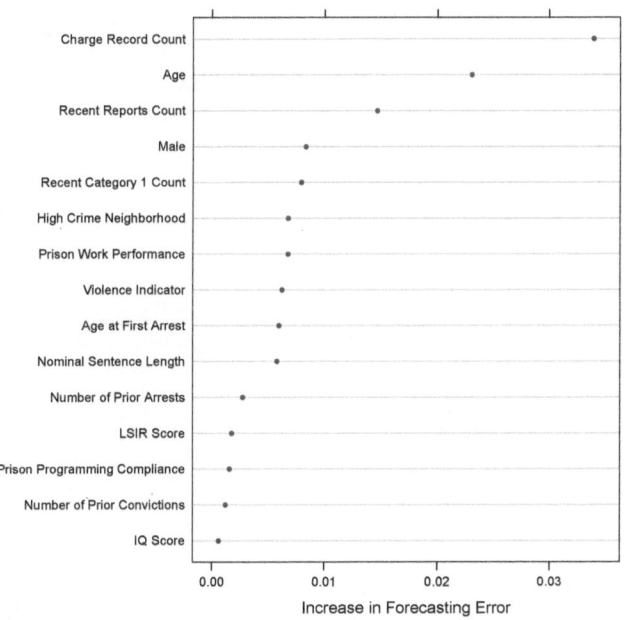

Fig. 2.1 Predictor importance measured by proportional reductions in forecasting accuracy for violent crimes committed within 2 years of release.

Figure 2.1 shows the results when an arrest for a violent crime (called here a "Level 2" crime) is being forecasted. Associated with each predictor is the reduction in forecasting accuracy when that predictor is not allowed to contribute. For

---

largely for the young. Is that theory? Likewise, anyone who had analyzed data on the biographical factors related to violent crime has likely found the same thing. Is that theory? And even when criminology theory is reviewed specifically for the role of age, the mechanisms by which individuals "age out" of crime are unclear. Everything from changing concentrations of certain steroidal hormones to marriage has been proposed. Is that theory? There is no doubt that as an empirical matter, age is related to crime. A forecasting procedure finding otherwise surely needs to be very carefully scrutinized. If there really is theory to help, all the better. The point is that far too often claims said to be based on subject-matter theory are really not.

example, when the number of prison misconduct charges is excluded, forecasting accuracy drops about 4 percentage points (e.g., from 60% accurate to 56% accurate). It is the most potent predictor of the group. Because the predictors are often substantially related to one another, there is also overlapping forecasting power that cannot be unpacked. Still, it is clear that many of the usual predictors surface, with perhaps a few surprises. The role of behavior in prison is far more important than prior record, and the popular LSIR score contributes little beyond what the other predictors bring.

Output describing *how* predictors are related to the response is also important. For example, the use of race as a predictor is quite properly a sensitive matter and often ruled out *a priori*. But then, one should know the manner in which potential race surrogates such as neighborhood of residence are related to re-offending. How predictors are related to the response can also affect the ways in which the forecasts are used in practice. There will be, for instance, circumstances in which the forecasts will be made with simple pencil-and-paper check lists. How these instruments are constructed will depend on information beyond a predictor's forecasting importance.

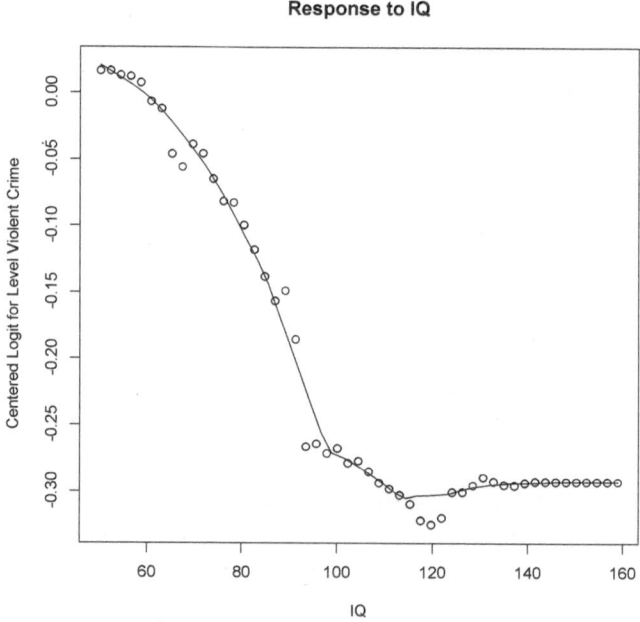

Fig. 2.2 How inmate IQ is related to whether a violent crime is committed while on parole. The relationship is negative for below average IQ scores and flat thereafter.

To illustrate, Figure 2.2 is a plot of how IQ is related to commission of a violent crime while on parole. Ignoring the units on the vertical axis for now — a larger value means greater risk — it is apparent that IQ as measured in prison is related to the risk of being arrested for a violent crime. A strong, negative relationship holds for IQ scores between about 50 and 100. For higher IQ scores, there is no relationship. IQ has some predictive value only for inmates with below average scores. It matters if an inmate has an IQ of 70 compared to an IQ of 90. It does not matter if an inmate has an IQ of 110 compared to an IQ of 130.

If the point is inform decision-makers in real time, the forecasting procedure must be capable of performing properly in the real-world settings in which those decisions are made. Good forecasting procedures can easily be ported to agency servers or even laptop computers so that data input, data processing, and data output require little more than a few keystrokes. There can also be circumstances when computer-based forecasts are impractical. For example, forecasts can be used to help inform police decisions in the field. As just noted, a simple check list of risk factors may then be appropriate (*Berk et al., 2005*). A good forecasting procedure must provide an effective way to arrive at an adequate paper-and-pencil format.

All forecasts come with uncertainty. If decision-makers are to make the best use of criminal justice forecasts, that uncertainty should be quantitatively represented in a form that is easy to understand. Forecasting error bands are an illustration. However, forecasting uncertainty can come from many different sources: random sampling of the data, measurement error, modeling errors, the forecasting procedure itself, and others. As a technical matter, it is very difficult to properly take all of the sources of uncertainty into account. Still, first approximations of uncertainty can be terribly important and for some forecasting procedures, relatively easy to obtain.

Finally, some will claim that the forecasting procedures should capitalize on social science theory. This is certainly a good idea when that theory can be helpful. For example, recent work by Blumstein and Nakamura (*2009*) has documented the important role of "redemption." An important predictor of reoffending on probation can be the elapsed time between the present and most recent past arrest. The longer that interval, the less the risk of re-offending. If the elapsed time is long (e.g., a decade) the risk of committing a serious crime can be as low, or even lower, than the risk in the general population. But social science theory can also be an unnecessary straightjacket. On one hand, good theory can help determine which kinds of predictors could be useful. On the other hand, there is the danger of overlooking predictors that social science theory has not recognized. Of late, this has become a serious issue because data mining approaches have been uncovering useful predictors and relationship previously overlooked, or at least not taken seriously. For example, an armed robbery committed at age 16 can be powerful predictor of future violent crimes. The same armed robbery committed at age 35 will likely have little predictive value. Age and prior record individually can be less predictive than when used in combination. In short, it is helpful to have a forecasting procedure that can inductively arrive at an effective set of predictors from a very large pool of candidates and not be constrained by prevailing social science theory. Known risk factors will likely be rediscovered and new risk factors can be unearthed — a win-win situation.

Chapter 3
A Conceptual Introduction to Classification and Forecasting

Abstract Because the criminal justice outcomes to be forecast are usually categorical (e.g., fail or not), this chapter considers crime forecasting as a classification problem. The goal is to assign classes to cases. There may be two classes or more than two. Machine learning is broadly considered before turning to random forests as the preferred forecasting tool. The approach is conceptual rather than formal. Some readers may find the material challenging, but the stage is being set for demanding material in chapters to come.

3.1 Populations and Samples

In this chapter, the emphasis is on characterizing patterns in a given dataset. However, the underlying premise, more central to later discussions, is that the dataset is a sample from a population of policy interest. Forecasting essentially requires this perspective because as already discussed, cases for which forecasts are needed are properly assumed to come from that same population. A key implication is that inferences are necessarily being drawn from the dataset to the population, or at least certain elements of the population defined by the values of predictors. This, in turn, raises the key question of *how* the data were generated. We leave that discussion for the next chapter. For now, it will suffice to assume that the dataset is effectively, if not literally, a simple random sample from an appropriate population.

3.2 Classification and Forecasting Using Decision Boundaries

Classification is a process of assigning classes to objects. Here, that means assigning particular crime categories to individuals. The individuals are people who have run afoul of the criminal justice system and have been at minimum arrested. Many are then processed by the courts, prisons, and parole/probation departments. At each step, criminal justice personnel make decisions about how best to proceed. These decisions are often informed by the risk to public safety an individual presents. The assessment of such risks depends on assigning a crime class, which can serve as a forecast, to each individual. An individual who is classified as a "bad guy" is subsequently treated as a "bad guy." An individual who is classified as a "good guy" is subsequently treated as a "good guy."

Two steps are required. First, an actuarial method is used to characterize how various properties of individuals and their immediate crimes (e.g., the crime for which they were just arrested) are associated with different kinds of subsequent outcomes (e.g., an arrest). Second, with those relationships established, the associations can be used to place new individuals into a crime class when the crimes they may commit are unknown. To take a simple example, if men under the age of 21, who are gang members, who have several past arrests for serious crimes, and who first appeared in court before the age of 14 commonly fail on probation through crimes of violence, all individuals with that profile can be classified as threats to public safety before they have an opportunity to reoffend.

Classification Using A Decision Boundary

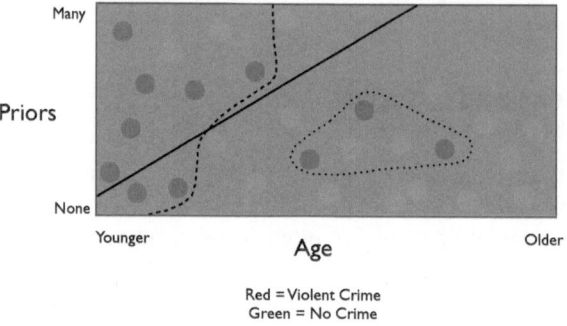

Fig. 3.1 Classification by two crime outcomes using the number of priors and years of age. Linear boundaries are used.

Behind such reasoning are several important issues, some of which can be subtle. Figure 3.1 provides a visual aide. The colored circles represent individuals. The color indicates the observed outcome: red for an arrest for violent crime and green for all other outcomes. The outcome is binary, and that is the key point. Any other two-category outcome would do for now.

There are two predictors: age and the number of prior arrests. Age ranges from younger (e.g., 18 years old) to older (e.g., 60 years old). The number of priors ranges from none (i.e., 0) to many (e.g., 50). The variables are just meant to be illustrative.

Figure 3.1 can be interpreted as a 3-dimensional scatter plot. Thus, the red circle in the upper left hand corner is an individual arrested for violent crime, who has many priors and is young. The green circle in the lower right hand corner is an individual not arrested for a violent crime, who has very few priors and who is older. All of the other circles have analogous interpretations, and generalizations to datasets with more than three predictors employ similar reasoning.

The goal of classification is to subset the data by predictor values so that the classifications are as accurate as possible. Consider for now just the straight solid line, called a "linear decision boundary." This linear decision boundary could be constructed from least squares regression with the outcome coded as "1" for a violent crime and "0" for all other outcomes. Predictors would be priors and age. The line would be drawn for fitted values of .50. All observations above the decision boundary would have fitted values greater than .50. All observations below the decision boundary would have fitted values equal to or less than .50. Individuals above the decision boundary would tend to be younger and have a greater number of priors. Individuals on or below the decision boundary would tend be older and have fewer priors. The term "decision boundary" is used because, in effect, the location of the boundary determines how cases are to be classified.

The fitted values can be interpreted as conditional proportions. For each configuration of age and prior record, the fitted value is, according to the regression model, the proportion who commit violent crime. If the data are seen as a realization of a stochastic process represented by the regression model, such proportions can be interpreted as probabilities. Probability interpretations are common, although too often unjustified. The connection between the regression model and assumed stochastic process is obscure.

How accurate is the classification? The subset to the upper left of the decision boundary contains 6 individuals who committed a violent crime and 3 who did not. One might label this subset as "violent" using a majority vote criterion. There are 6 votes for violent crime and 3 votes for other outcomes. Individuals in this data subset are twice as likely to commit a violent crime compared to all other outcomes. One might classify all individual who fell in this region as "violent." Then, there would be 3 classification errors.

The subset to the lower right of the decision boundary contains 14 individuals who did not commit a violent crime and 5 individuals who did. By the same sort of majority vote reasoning, one might classify any individuals who fell in this region as nonviolent. There would then be 5 classification errors.

There are 3 classification errors in the class of violent offenders and 5 classification errors in other class, for a total of 8. In principle, the number of forecasting errors could be reduced if the decision boundary were allowed to be nonlinear. Such a boundary could be more responsive to patterns in the data. The dashed line is an example that could result for including quadratic and cubic functions of the two predictors in the regression. If the dashed line instead of the solid line is used as the decision boundary, the number of classification errors is reduced to 6.

There can also be decision boundaries defining regions that are not contiguous. The dashed line and the dotted line define two such subsets of the data. This might be the result of including product variable for age and priors in the regression. By majority vote, individuals in either region would be classified as violent. There are now 4 classification errors.

But what use is the class assigned to a region of the scatter plot? The outcomes are already known. We can simply classify each individual by the observed outcome with no need of predictors whatsoever. Classification would then be perfect. There would be no errors.

The class labels are essential for forecasting. Any subsequent cases not in the training data and for which the outcome is unknown can be subject to a forecast that would depend on which side of a decision boundary they fall. In the linear case, falling above the boundary forecasts a violent crime. Falling on or below the boundary forecasts an absence of violent crime. In short, the point of assigning classes to subsets of the data is to use those classes for forecasting.

In Figure 3.1, classification is undertaken with least squares regression and a binary 1/0 coded response variable. For a variety of reasons, this is not an ideal approach. One can have, for instance, fitted values less than 0 and/or greater than 1.0 suggesting some fundamental error in the regression model. Simple and better alternatives could be logistic regression or its close cousin, linear discriminant function analysis. There are also powerful machine learning approaches such as support vector machines, boosting and Bayesian trees that can be used for classification. Accessible and far more complete textbook discussions can be found elsewhere (*Berk, 2008b; Hastie et al., 2009*). For reasons that will soon be clear, we favor another method to arrive at decision boundaries.

One other preliminary point needs to be made. Thus far, we have classified by majority vote, giving each observation the same voting weight. Likewise, when we evaluated classification accuracy, we treated all classification errors the same. A "red" classified as a "green" counted the same as a "green" classified as a "red." But, as discussed previously, perhaps when an individual who committed a violent crime is classified as being violence-free, a potentially serious error has been made. This error may be more serious than classifying a person who was violence-free as committing a violent crime. As emphasized earlier, not all forecasting errors are the same. We address this issue in more depth below

3.3 Classification by Data Partitions

One of the key features of the statistical tools emphasized here is the ability to inductively find unexpected and often highly nonlinear relationships between predictors and an outcome to be forecasted. An important device to this end is breaking up quantitative predictors into sets of indicator variables. Figure 3.2 provides a simple illustration for a single predictor.

Figure 3.2 plots the proportion of parolees who fail while under supervision against their ages. The proportions are shown with the gray circles. The proportion who fail decreases rapidly from about age 18 to 21, more slowly from 22 to 30,

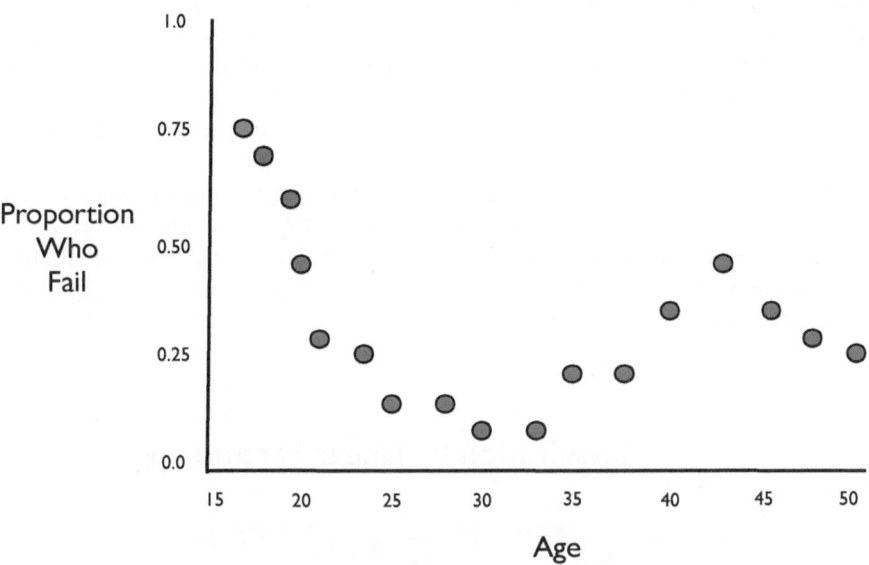

Fig. 3.2 Plotting the proportion who fail by age, breaking up age into a set of indicator variables. Each age gets its own indicator variable. The nonlinear relationship is apparent and empirically ascertained.

increase until the middle 40s, and decline slowly up to age 50. We will see patterns rather like this later when real data are examined.

It is unlikely that a nonlinear relationship between failing on parole and age would have been anticipated with any real precision, and even more unlikely that an appropriate functional form would have been specified *a priori*. But if each value of age were represented by an indicator variable, one could let the data empirically approximate the requisite functional form.

Indicator variables, also called "dummy variables," are constructed in the same fashion. An indicator variable is generally coded equal to "1" if some condition is met (e.g., age $= 21$) and equal to "0" otherwise (e.g., age $\neq 21$). Regressing whether or not an individual failed on the age indicator variables would have produced a graph of the fitted values much like Figure 3.2.[1] The relationship between failing and age would then be visually apparent.

If instead of constructing an indicator variable for each value of age, there were an indicator for each of a set of age ranges (e.g., 18-21, 22-25, etc.), there would have been fewer fitted values and a less detailed rendering would have resulted. But in trade, each indicator variable would be constructed from a larger number of observations. This could, in turn, increase the stability of the estimates. We will revisit this issue later. The general point for now is that by discretizing quantitative predictors, one can often empirically characterize previously unknown and highly nonlinear associations.

We now return to a 3-dimensional scatter plot. In Figure 3.3, the predictors are the same as before, but there are now three crime outcomes: red for violent crime, yellow for nonviolent crime, and green for no crime. Two crime classes could have been used; however, with three some new issues are arise that are especially important for the applications to be discussed later.

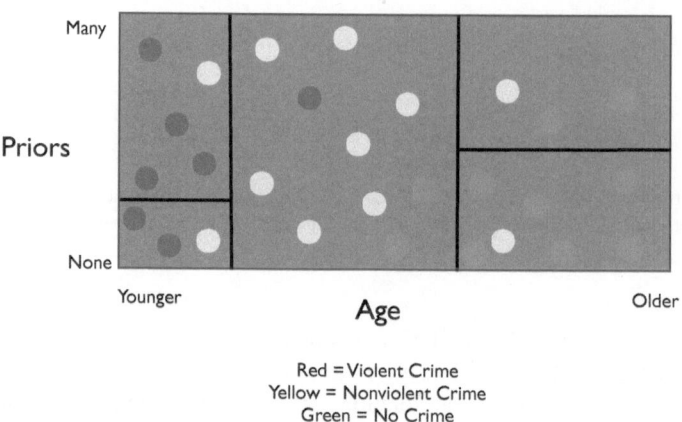

Classification by Linear Partitioning

Red = Violent Crime
Yellow = Nonviolent Crime
Green = No Crime

Fig. 3.3 Partitioning three crime outcomes by the number of priors and years of age. Linear boundaries are used

---

[1] An indicator variable implies that the relationship with the response variable is a step function. The regression function, therefore, is a linear combination of step functions. If an intercept is, as usual, included in the regression, one of the indicators would need to be dropped. Otherwise, the regressor cross-product matrix would be singular, and there would be no unique solution from which to obtain estimates of the regression coefficients.

As before, there are decision boundaries defining partitions of the data. In formal language, they are called hyper-rectangles. There are now five such partitions defined by straight lines that are parallel to one of the two axes: "axis-parallel linear splits" (*Ho, 1998: 833*). As before, each partition has its location determined by the age and prior record of individuals in the partition. Individuals in the upper left partition, for instance, are once again younger and have longer prior records. Individuals in the lower right partition are once again older and have shorter prior records. The other partitions can be characterized in a similar fashion.

Each partition can be represented by an indicator variable defined by partition boundaries. Suppose that for the upper left partition, the horizontal boundary fell at a priors value of 3, and the vertical boundary fell at an age value of 20. Then the upper left partition could be represented by an indicator variable coded "1" for cases with 3 priors or more and with ages less than 20, and "0" otherwise. Indicator variables could be defined in a similar fashion for all other partitions. This is just a generalization of the earlier discussion of discretizing quantitative predictors. Each indicator variable is a function of two predictors rather than one. Again, the goal is to let the data determine the ways in which the predictors are related to the response variable. If there are substantial nonlinearities, they are likely to be found. That depends on how the decision boundaries are established, which we consider soon.

It should be clear that younger individuals with longer prior records are more likely to be in partitions where the most common outcome is violent crime. Likewise, older individuals with shorter prior records are more likely to be in partitions where the most common outcome is no crime. And individuals who are in the middle ranges of age and prior record are more likely to be in partitions where the most common outcome is nonviolent crime. But no particular functions of the age and prior record are imposed. There is nothing like the usual slopes assumed by linear regression.

3.4 Forecasting by Data Partitions

How can such information be used for forecasting when the outcome has not yet been observed? We proceed in the same spirit as earlier. Consider the partition at the upper left. Four of the five individuals have committed a violent crime. One of the five committed a crime but not a violent one. If one wanted to assign a crime class to the partition as a whole, what would that crime class be? Because four of the five individuals in the partition committed a violent crime, one might reasonably decide to attach the violent crime label.

The same reasoning can be applied to each of the partitions shown. Each partition can be labeled according to the largest fraction of individuals with a given outcome. The outcome with the most "votes" determines each partition's label. For example, the large middle partition would be assigned a nonviolent crime label because seven of the ten individuals in that partition committed a nonviolent crime, one of the ten committed a violent crime, and two of the ten committed no crime.

As before, forecasting capitalizes on the labels. New individuals for whom the outcome is unknown are assigned to the partition corresponding to their ages and prior record. Once assigned to a partition, the crime label attached to that partition is applied. Thus, all individuals assigned to the upper left partition are projected as violent crime risks. All individuals in the large middle partition are projected as nonviolent crime risks. Each partition would be used in a similar fashion.

3.5 Finding Good Data Partitions

So far, the locations of the partitions have been taken as given. In practice they are not. How best to place the partition boundaries is the main task of any method used to arrive at a forecasting procedure. Ideally, the boundaries should segment the scatter plot so that once the class assigned to each partition is determined, the number of misclassifications, or some function of them, is as small as possible. In Figure 3.3, the upper left partition, for instance, has one classification error. The large middle partition has three classification errors. Over all partitions, there seven classification errors. One could in principle do better or worse.

There is, of course, a trivial and useless solution in which each partition contains only one individual. It is useless because there is no role for any predictors. Each individual is treated as unique, and there is no principled way to construct forecasts for new individuals for whom the outcome is unknown. For the moment, therefore, assume the number of partitions has been determined and as a result, there is a substantial number of individuals in each. Also for the moment assume that the boundaries are straight lines. How then are the partition locations determined?

There is alway a brute force solution. One can try all possible sets of, say, 5 partitions and pick the set for which the number of classification errors, or some function thereof, is as small as possible. A key advantage of this approach is that one can proceed in a highly inductive manner that can be extremely effective when how best to partition the data is not known *a priori*. The data analyst goes wherever the data lead. Then, with the best set of partitions in hand, one can assign a class to each partition by counting votes, and classifying all individuals in a partition accordingly.

In most real situations, a brute force approach is not practical. The most apparent obstacle is the very large number of possible partitionings of the data. Even with a powerful computer, the time constraint may be binding for even a moderately sized training sample. There are several options. Perhaps the most common approach, and the one on which we will build, is to proceed sequentially.

This is the basic idea. Another way of thinking about the data partitions is that ideally, the observations within each partition should be as homogeneous as possible with respect to the outcome variable. The best result would be a set of partitions in which each contained individuals with the the same outcome: all committed a violent crime, all committed a nonviolent crime, or all committed no crime. In Figure 3.3, each partition would contain circles of a single color. Such within-partition homogeneity is ideal because the classification is then perfectly accurate. There are no classification errors.

The worst result would be a set of partitions in which all three crime outcomes were equally represented. A third of the circles would be red, a third would be yellow, and a third would be green. Within-partition heterogeneity would be as large as possible. No outcome was more common than any other. Then, a one-observation-one vote cannot be used to determine the class assigned, and classification would be as inaccurate as possible no matter which class were attached.

The partitioning goal, therefore, is to get as close to perfect homogeneity as possible. Starting with no partitions, a single boundary is selected that reduces heterogeneity the most. That is, the two partitions that result should produce the most homogeneity possible for any two partitions of the data. The same approach is then applied separately to the two partitions. Now there are four partitions, and each of the four is separately split as before. The process continues until the predetermined number of partitions is reached or a partition contains the minimum number of observations allowed. A more formal and complete discussion of the process is provided later.

The sequential approach is sometimes called "recursive partitioning." Recursive partitioning is often characterized as "greedy" because at each step the best outcome is selected, which is not then subject to revision in later steps. Recursive partitioning usually does not lead to the most accurate classification possible, but generally does quite well and can be very fast. Other kinds of "greedy" approaches are popular for similar reasons.

3.6 Enter Asymmetric Costs

In our discussion so far, classes have been assigned to subsets of the data by vote. For a given subset, the class with the most votes is the class assigned. All observations not of that class are then classification errors. One happy result is that the number of classification errors is minimized. Assigning any other class would produce more classification errors.

But this rationale rests on a very important assumption: the vote of each observation counts the same and therefore, all classification errors are equal. In practice, the one-observation-one-vote requirement can be a substantial problem. Consider again Figure 3.3. The partition in the upper right corner is assigned the class of no crime by a vote of 2 to 1. There is then a single classification error. An individual who was arrested for a nonviolent crime is classified as if no crime were committed.

Had the class of a nonviolent crime been assigned to the partition, two individuals who committed no crime would be classified as if they had each committed a nonviolent crime. Is that really worse? It depends on how one weights the classification errors and that depends on how those classifications are used.

As before, suppose classification is used in forecasting. If the values of the two predictors place an individual in the upper right data partition, the partition class becomes a forecast. The forecasts are meant to inform decisions. For a forecast of no crime, the individual is released on parole. For a forecast of a nonviolent crime, the individual is denied parole.

Both decisions can be wrong. The individual released may be arrested for, say, burglary. The individual held in prison might have been crime free if released. A variety of costs can be associated with both forecasting errors.

Suppose the person incorrectly released on parole is arrested for drug trafficking. The costs should include, at least, the costs to law enforcement for apprehension and processing as well as any consequences for the neighborhoods in which the drug trafficking was undertaken: violent disputes over turf, crimes committed to obtain money with which to buy drugs, health implications for local drug users, the drain on drug treatment programs, and fear of crime for law-abiding residents.

The costs associated with holding an individual who would have been crime free include the negative emotional experience of additional time in prison, damage to the individual's human capital and later adjustment potential after release, the cost in tax payer dollars of prison housing and supervision, and any destructive impact on the individual's family and larger neighborhood.

Because some forecasting errors are more costly than others, it makes sense that these costs should somehow figure in to the forecasting enterprise. For example, if the costs for failing to incapacitate are more than twice as large as the costs of unnecessary incapacitation, the class assigned to the upper left partition should be nonviolent crime, not the absence of crime. The total costs of the forecasting errors would be less than had the original class been used.

Consider a very simple illustration. Suppose the current class of *no crime* were assigned. There is a single classification error. It has some cost, say, C. Now suppose that the *nonviolent* crime class were assigned. There are two classification errors, each with a cost that is, say, one-third the size: $1/3 \times C$. One should assign the class having the lower costs for classification errors.

If *nonviolent* crime is the assigned class, the cost of forecasting errors is $2 \times (1/3 \times C) = \frac{2}{3}C$. If *no crime* is the assigned class, the cost is C. By this reasoning, the class of *nonviolent crime* should be assigned, *not* the class of *no crime*. *The proper forecast changes*: now, all individuals falling in that partition would be classified as committing nonviolent crimes, and any forecasts would be for nonviolent crimes as well.

When the costs of all forecasting errors as assumed to be the same, one is said to be working with "symmetric" costs. Under symmetric costs, a simple vote properly determines the class assigned to a data partition. When the costs forecasting errors are assumed to be different, one is said to be working with "asymmetric" costs. In this case, the votes need to be weighted by their *relative* costs. In effect, a different "winning" threshold is being imposed (*Monahan and Solver, 2003*)

In summary, forecasts of criminal behavior commonly ignore the relative costs of different kinds of forecasting errors. When the costs of forecasting errors are not the same, forecasting errors must be weighted accordingly. These weights can then cascade through the actuarial methods with a range of consequences for the output. Perhaps most importantly, the forecasts will be affected, often substantially. When asymmetric costs are ignored, the forecasts are not likely to be responsive to the needs of decision makers, who can be badly misled.

3.7 Recursive Partitioning Classification Trees

When data partitioning is done in a recursive manner, the results are often shown as a classification tree.[2] Figure 3.4 is a classification tree for a recursive partitioning of Figure 3.3. The sequence of splits is in this instance illustrative. The observations shown are too few to do a meaningful analysis.

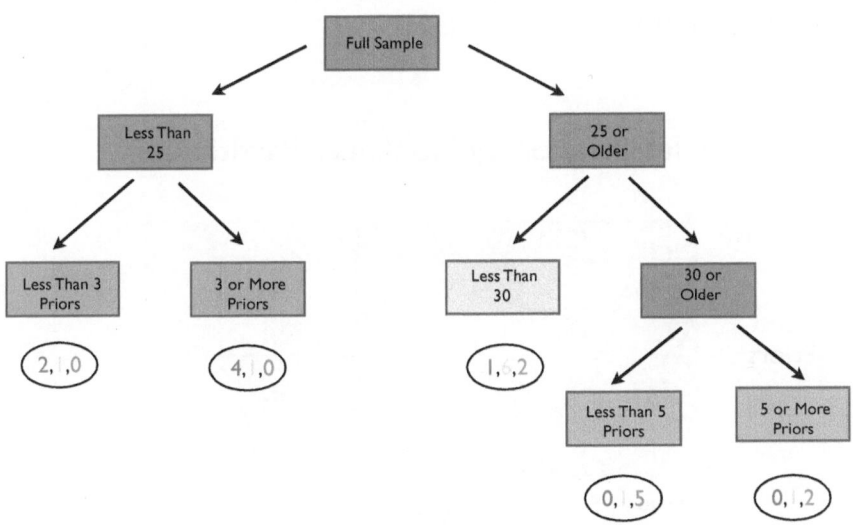

Fig. 3.4 A classification tree with two predictors and three outcome categories. Below each terminal node is a count of the number of cases falling in each class (Red = violent, yellow = nonviolent, green = no crime). The terminal nodes are color-coded to show the class assigned by plurality vote under symmetric costs.

---

[2] Classification trees is a special case of classification and regression trees (CART). A regression tree uses recursive partitioning with a quantitative response variable. A classification tree uses recursive partitioning with a categorical response variable.

Each rectangle represents a subset of the data (except the one at the top which represents all of the data), and is called a "node." The arrows show which cases go where. For example, under the first split of the data, all individuals with ages less than 25 are placed in the left partition. The nodes at the bottom are called "terminal nodes." It is these nodes to which crime labels are assigned. A simple plurality vote is applied here. If the costs of forecasting errors are asymmetric, the votes can properly be weighted by relative costs. Such weighting is certainly a good start but by itself insufficient. The weighting should be introduced in a fashion that affects the way the splits are determined as well. The result can then be trees with different structures. For example, the first break might be for prior record at a value of 2 or greater, which changes the composition of all subsequent nodes. We return to this matter in the next chapter.

The votes are shown in the ellipses below each terminal node, and the class assigned to each terminal node is also color-coded. When used for forecasting, each case is paced in a terminal node based on the values of its predictors (e.g., 18 years old with 4 priors) and assigned the projected outcome class represented here by the color of the terminal node.

Classification by Nonlinear Partitioning

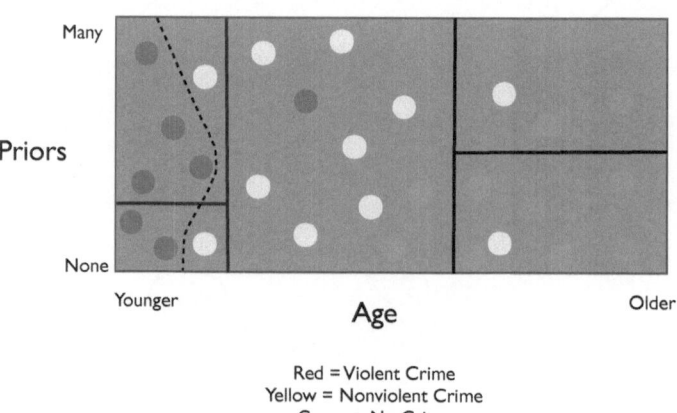

Red = Violent Crime
Yellow = Nonviolent Crime
Green = No Crime

Fig. 3.5 Partitioning three crime outcomes by the number of priors and years of age. nonlinear boundaries are now considered, and classification accuracy improves.

In practice, one would start with the full training sample. The tree-construction goal is to partition the data so that the partitions are as homogeneous as possible — a formal treatment will follow soon. Suppose the first boundary is constructed from age. Those under 25 years of age are sent down the left path and those 25 and older are sent down the right path. This corresponds to the segments on either side of the left most vertical line in Figure 3.3. Next, suppose the left-most partition is subdivided by the number of priors: less than 3, and 3 or more. The initial right age partition might be next segmented by age again: below 30, and 30 years of age or more. Finally, the oldest age group could be segmented by the number of priors: less than 5, and 5 or more. The last set of partitions corresponds to the terminal nodes.

One does not have to be limited to straight line boundaries. nonlinear boundaries are far more flexible than linear boundaries and in principle, can be used to build partitions of the data that have fewer classification errors. Figure 3.5 is the same as Figure 3.3 except that the two blue linear boundaries on the left can be replaced by a single nonlinear boundary shown with the dashed line. Two classification errors have been eliminated so that the resulting partition on the far left contains only individuals who committed violent crimes. The other partitions could be improved as well with nonlinear boundaries. But it will often turn out that linear boundaries are easier to implement and perform surprisingly well.

3.7.1 How Many Terminal Nodes?

More needs to be said about an appropriate number of terminal nodes. For a fixed number of observations, a greater number of terminal nodes means that on average, each node will contain fewer observations. A key implication is that if there is a desire to undertake statistical inference from training data, the class assigned to each partition is determined by a smaller sample size. And recall, that inferences beyond the data on hand are implied by all forecasting.

Returning to an earlier example, suppose a terminal node includes only men under the age of 21, who are gang members, have several past arrests for serious crimes, and first appeared in court before the age of 14. Suppose also that there are only 5 such individuals in the training data and that there are 3 votes for a violent crime and 2 votes for a nonviolent crime. The proportion committing a violent crime is .60. Consequently, the violent crime category is attached to all subsequent individuals who have the same profile, and serves as a forecast.

The intention to use the assigned class for forecasting implies that there are individuals who are not included in the training data or test data, who will be treated as members of the same population from which the training data and test data were taken. That is, the data used to build the forecasting procedure are but one of many possible realizations (or samples) from a population. Therefore, the proportion of .60 is really an estimate of a population parameter, and needs to be treated as such. Estimation bias and sampling error should be considered.

In this example, the number of observations is very small. With so small a sample, the sampling error in the computed proportion of .60 is likely be substantial. In another random realization of the data, the proportion could easily turn out quite differently and another class might be assigned. In short, with a smaller number of observations in a terminal node (and hence, data partition), there is more sampling error; the class assigned and any forecasts based on that class will be more unstable. The variance in the vote proportion and class assigned will be greater.

But that is not all bad. For a fixed number of observations, when on average there are a smaller number of observations in each terminal node, there are more terminal nodes. This allows for more specialized data partitions. For example, one might be able to usefully distinguish between individuals with a 10th grade education and individuals with a 12th grade education when all other predictor values are otherwise the same. Put another way, if small differences in offender profiles really matter systematically when behavior on parole is forecasted, having more terminal nodes is good. Forecasting bias is reduced.[3]

Clearly, there is a tradeoff between the variance and the bias in terminal node forecasts. For a given number of training observations, a larger number of terminal nodes will tend to reduce the bias and increase the variance. A smaller number of terminal nodes will tend to increase the bias and reduce the variance. This is an example of the *bias-variance tradeoff* about which we will have a lot more to say later. Ideally, it is possible for any given training data to find an appropriate balance between the bias and the variance so that on the average the class forecasted is correct as often as possible.

3.7.2 Classification Tree Instability

The bias-variance tradeoff can apply to the entire tree, not just the classes assigned to the terminal nodes. But because greedy algorithms can lead to unstable results, the more pressing problem is usually the variance, not the bias. Any subsetting decision made earlier in the process affects all later partitions. If the location of an initial partition is unstable, so are all of the other partitions.

Instability can result if there is more than one partition of the data with nearly the same improvements in partition homogeneity. Over realizations of the data, different initial partitions can be selected by chance alone, because small differences in the composition of each sample translate into different initial partitions. In Figure 3.4, for instance, the first partition might be based on the number of priors rather than age. Subsequent partitions would then likely differ as well. The same reasoning applies to any other partition before the terminal nodes.

However, the instability matters a lot more for some applications than others. The sequence of partitions can make a very important difference in any *interpretations* of the relationships between the predictors and the response. For example, from Figure 3.4 one might want to claim that individuals under 25 years of age present especially grave threats to public safety. 25 years of age was the initial split that reduced within-partition heterogeneity most. Some might even claim that age should

---

[3] In this instance, "bias" refer to a systematic tendency to underestimate or overestimate the terminal node proportions in the population responsible for the data. A more formal treatment follows in later chapters.

be seen as the most important factor in violent crime. But in another sample, the first partition might single out individuals with more than 3 prior arrests. Some might then say that prior record is more important than age.

Instability can be less problematic for forecasting. What matters are the terminal nodes. The path by which one arrives at the terminal nodes does not matter. So, if a terminal node is defined by young individuals with long prior records, it does not matter whether age or prior record was chosen first to subset the data. The forecasts are the same. As a result, instability has less impact; however, it should not be ignored.

There are several good ways to address instability. Some attack the problem directly. Perhaps the earliest and best known example is "bagging" (*Breiman, 1996*). The basic idea is to draw a substantial number of random samples from the training data. Each sample usually has the same number of observations as the training data. By sampling rows of the data *with replacement*, each sample will almost certainly differ from one another and from the training data. Recursive partitioning is then applied each sample. At the very least, it is then possible see how the tree diagrams can vary from random sampling alone. More stable classification and forecasts are then obtained by averaging over samples. An individual is classified or forecasted by plurality of votes over trees. If there are, say, 150 trees, an individual can be classified or forecasted up to 150 times. Whichever class gets the most votes is the chosen class. The average class is likely to be far more stable than the class assigned from a single tree. Averaging cancels out much of the instability. But there is a price. There is no longer a single tree structure to interpret and no apparent way to usefully define, let alone construct, an "average tree."

There are other approaches. For example, "Boosting" (*Friedman, 2002*) applied to classification trees also makes many passes through the data and classifies each case each time. But there need be no sampling. With each pass, the data are reweighted so that cases that are more difficult to classify correctly are given more weight. The final class for each case is again determined by a plurality of votes, but votes from passes that classified more accurately are counted more heavily. The intent of the weighting is to extract more information from the available predictors, but as a byproduct, stability can be improved.

In this book, we favor random forests, which can be seen as an extension of bagging. A large number of trees is constructed from samples of the training data drawn with replacement. In addition, when any split for any tree is considered, only a random sample of predictors is evaluated to arrive at the best partitions. As with bagging, classes are assigned by a vote over trees. Random forests has several other important features and will be discussed in depth in the next chapter.

Chapter 4
A More Formal Treatment of Classification and Forecasting

Abstract This chapter covers much of the foundational material from the last chapter but more formally and in more detail. The chapter opens with a discussion of the model used to characterize how the data were generated. That model is very different from the one used in conventional regression. Then, classification is considered followed by the estimation issues it raises. The bias-variance tradeoff is front and center. Some material in this chapter may require careful reading.

4.1 Introduction

The last chapter informally raised a number of major conceptual issues. We return now to those issues, and introduce others, but in a far more formal framework. The formality is necessary for clear exposition and because some matters can only be properly considered in that manner. The formality will also establish links to theoretical statistics and to a range of related concerns. This is not academic fluff. If one intends to usefully inform criminal justice decisions, there must be coherent and internally consistent reasoning linking the statistical methods used to the forecasts produced.

Much of what follows derives from a "thought experiment" by which the data on hand, and future data, are generated. There needs to be an account of how the data came to be that gives meaning to the mathematics employed. Without the structure an appropriate thought experiment provides, the mathematics is nothing more than manipulating symbols. One must keep the thought experiment in mind for the abstractions in the next few pages to make sense.

4.2 Data Generation Models

We begin with the thought experiment. Imagine that the N observations in the training data are N random realizations from a joint probability distribution, represented by $\Pr(G,X)$. G denotes the response variable, and X denotes the predictors. Both are random variables. Consequently, all of the data for each case (e.g., arrested while on probation, age, gender, number of prior convictions, etc.) are a realization from that joint distribution, and there can be in principle a limitless number of independent realizations for each observational unit.[1]

There is really nothing novel in the joint probability distribution formulation from a statistical point of view. One can find it discussed in any number of statistics textbooks (e.g., *Rice, 2006*).[2] It is also a common framework for machine learning of the sort considered in this book (*Hastie et al., 2009: 18*).

Some features of $\Pr(G,X)$ are not of direct interest but can nevertheless matter in principle. For example, it can be important for the distribution around the conditional exectation for each case to have the same variance. Note that this is a feature of the joint distribution, not the realized dataset. But for the machine learning applications considered here, such features can usually be ignored. We will not be considering the kinds of statistical inference for which they are likely to be important.

No other assumptions are made about how the data are generated. There is, for example, there is no causal machinery and no concern about cause and effect. There are no disturbances, which figure so centrally in causal models, and no such thing as omitted variables. Some important predictors may not be in the training and test data, but that only means that information that might be useful for forecasting is unavailable.

Some readers may be uncomfortable treating the data in hand as a realization from a joint distribution. But as a conceptual matter, forecasting only makes sense if the current observations and all future observations for which forecasts are needed can be seen as realizations of the same data generation mechanisms. Consequently, a data-generation model of some sort is required. Conventional causal models are very demanding in ways that are unnecessary for forecasting. The joint probability model of data generation provides a forecasting framework that justifies the formal mathematics with very little excess baggage. If one chooses to adopt neither a causal model nor a joint probability model, some other model must be provided.

In summary, for forecasting to make sense, the data must come from some appropriate population. This requirement was addressed at length earlier. We are adding

---

[1] One can be a bit more concrete by imagining that "nature" samples cases at random for a limitless population whose joint distribution is $\Pr(G,X)$. This is much like how one thinks about survey sampling for certain kinds of political polls, although then the population of, say, registered voters is finite, and a survey researcher does the sampling.

[2] For ease of exposition, we are at this point assuming the equivalent of simple random sampling by nature — each population unit has the same probability of selection and sampling proceeds without replacement. If appropriate for a particular application, more complicated sampling designs (e.g., stratified sampling) can be used instead.

at this point is that the data must be a *random* realization from that population. This is a very important foundation for much of what follows.

4.3 Notation

We use capital letters for the names of variables or sets of variables. They are not used when mathematical manipulations are required because then, more specificity is needed. For those operations, we use bold capital letter. Thus, $\mathbf{G}$ is $N \times 1$; there are N rows and one column of data. The p predictors will be collected in an $N \times p$ design matrix denoted by $\mathbf{X}$; there are N rows and p columns.

G is categorical with K categories, also called classes. As before, for example, K might be three with classes: committed a violent crime, committed a nonviolent crime, and committed no crime. The collection of possible classes is represented by C. One goal is to estimate G with $\hat{G}$ using information in the predictors, and for this we write $\hat{G}(X)$. These are the N estimated classes, one for each observation. Another goal is to then use the estimated classes as forecasts for new cases when the response is not yet known. One would proceed in essentially the same manner when the response is quantitative, but the analogous estimates would be of conditional means, not conditional proportions.

4.4 Classification

If we intend to classify as accurately as we can, we need a way to think about classification error. To begin, there is a $K \times K$ loss matrix L that has zeros along the main diagonal and nonnegative values everywhere else. The off-diagonal elements contain the losses from observations having a true class of C_k classified incorrectly as class C_l. An example is classifying an individual as committing no crime when that individual actually committed a violent crime. The off-diagonal cells will often contain the sum of such errors. Then, 1-0 loss is being applied. Each misclassification has a cost of one, and correct classifications have no cost. Other loss functions can be applied when the costs are not symmetric, and that is where we are going. But at the moment, 1-0 costs simplify the discussion with no important loss in generality.

Because the data are considered a random realization for which both the outcome and the predictors are random variables, we use as a performance measure the *expected* prediction error. Denoted by EPE, it takes the following form.

$$\text{EPE} = \text{E}[L(G, \hat{G}(X))]. \tag{4.1}$$

The two arguments in the loss function L are the actual responses and the predicted responses, the latter a function of X. The expectation is over realization of the joint distribution $\Pr(G, X)$. EPE is, in effect, the average loss as a function of the fitted classes and the actual classes, over realizations of the data. The expectation operator is worth emphasizing in a forecasting context. There will be random realizations of predictor values for which a response will need to be predicted. The goals is to be as accurate as possible on the average.

Equation 4.1 can be usefully unpacked as

$$\text{EPE} = \text{E}_x \sum_{k=1}^{K} [L(C_k, \hat{G}(X)] \text{Pr}(C_k|X). \tag{4.2}$$

To make this more concrete, suppose there are two parole outcome categories: fail or not. K is two, with C_1, say, as fail and C_2 as not fail. Suppose there are two predictors, age and gender, represented by X. The term on the far right, $\text{Pr}(C_k|X)$, is the conditional probability of a given outcome (e.g., fail) for a particular configuration of predictor values (e.g., 28 year old males). The term just to the left in square brackets, $[L(C_k, \hat{G}(X)]$, is the associated loss that depends on the assigned class, $\hat{G}(X)$ and the actual class, C_k. Whatever that loss, it is multiplied by the conditional probability of the given parole outcome given the values of X. Outcomes that are more common for given predictor values, are weighted more heavily when the loss is computed. One imagines these operations for each configuration of predictor values and for each, summing over both parole outcomes. However, X is random and we are interested the expected loss. Consequently, the expectation of the loss over X is required. We wind up with an "average" loss over the joint distribution of the predictors. Nothing is being said yet about how the estimated $\hat{G}(X)$ is produced.

We seek to minimize Equation 4.2, and can proceed conditioning on X as follows.[3]

$$\hat{G}(x) = \text{argmin}_{c \in C} \sum_{k=1}^{K} L(C_k, c) \text{Pr}(C_k|X = x), \tag{4.3}$$

where the job is to choose the class c from the options provided by C to minimize the EPE, and in that manner arrive at our estimates of the outcome $\hat{G}(x)$. With a 1-0 loss, Equation 4.3 simplifies to

$$\hat{G} = \max_{c \in C} \text{Pr}(c|X = x). \tag{4.4}$$

---

[3] Even though we begin with the joint distribution $\text{Pr}(G, X)$, it is sufficient to condition on X. This is why it is sometimes called a "point-by-point" approach.

Equation 4.4 defines the Bayes classifier.[4] We simply classify by the most probable class. We are basically back to the voting rule used in the last chapter. But now, we have a formal justification. We are minimizing expected prediction error. Moreover, as we saw in the last chapter, we can have a weighted voting procedure with the weights a function of costs that are not 1-0 or even symmetric. Perhaps the key take-away is that we want to assign classes so that *on the average* we do as well as we can. In practice, that is the very best we can hope for.

4.5 Estimation in the Real World

Our discussion of the Bayes classifier took the distribution of true classes, conditional on the predictors, as known. In practice, the distribution is unknown. Therefore, we need a good way to estimate that distribution, and that depends on having adequate actuarial tools. But, those tools will not perform as advertised if the data the tools require are not sufficiently complete.

There is no statistical solution for not having a reasonably complete set of predictors or at least all of the important ones. For criminal justice applications of sort we consider, the set of predictors will usually be substantially incomplete. We need, therefore, to abandon the goal of getting the assigned classes right on the average. That is, we are not drawing from the joint distribution $\Pr(G,X)$ responsible for the data, but from *another* joint distribution $\Pr(G,X^*)$ characterized by a subset of the predictors in X. This would be an omitted variable problem were we doing conventional causal modeling. In this setting, there formally is no such such thing, but we may systematically overestimate or underestimate, say, the parole risk for those convicted of murder randomly realized from the "true" joint distribution $\Pr(G,X)$.[5]

This is a game-changer. We no longer aspire to get the "right" classes for $\Pr(G,X)$. We aspire to do as well as we can by the criterion of expected prediction error while limited to data from $\Pr(G,X^*)$. These issues are a bit tricky. Let's tackle first bias in estimates of the conditional probabilities.

---

[4] The term "Bayes classifier" can be used because of formal connections to decision theory in statistics (*Rice, 2006: Section 15.2*). In an overstuffed nutshell, there is a decision to be made. In this application, the decision is what class to assign to a case. There is an uncertain state of nature. In this application, that uncertain state of nature is the true class. It is uncertain because the true class depends on predictors, which are random variables. There is a cost associated with the decision that is a function of the true class and the class assigned. We want a decision rule that will on average minimize that cost. The average is taken over the distribution of the predictors on which the true classes depend. It follows that Equation 4.1 is "Bayes risk" for classification. The reference to "Bayes" comes about because the true classes have some probability distribution, that here happens to depend on predictors. The true classes not fixed. Equation 4.4 minimizes Bayes risk and consequently can be called a "Bayes rule."

[5] Recall that there is no regression model and therefore, no disturbances that might include the impact of variables not in the model.

4.5.1 Estimation Bias

Consider Figure 4.1. The figure is not a conventional scatter plot. It is a way to visualize some important features of the requisite statistical theory. On the horizontal axis is a single predictor. We use one predictor to simplify the discussion, but the conclusions generalize to the multivariate case. Age is used because it is familiar and well known to be related to recidivism. On the vertical axis are the probabilities of failure on parole, given age. Failure is binary: fail or not. The blue circles are the conditional probabilities of failure for different ages. For example, for 21-year-olds, the probability of failure is about .50. The blue circles show the "true" probabilities in the sense that they represent the conditional probabilities derived from the joint distribution $\Pr(G,X)$.

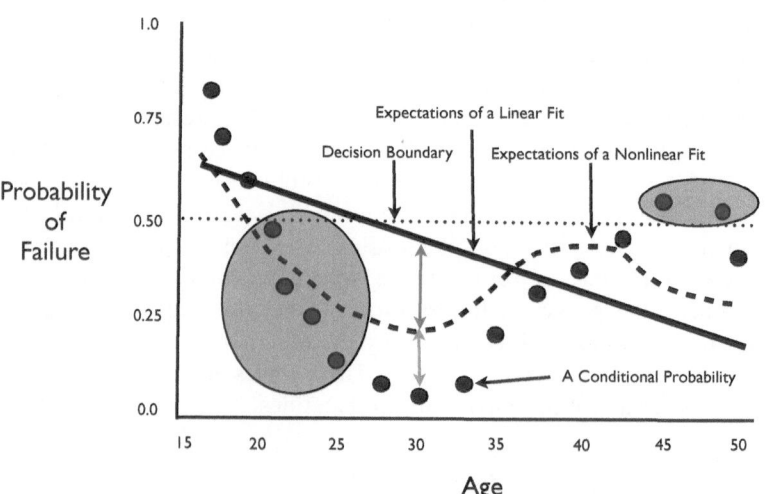

Fig. 4.1 A demonstration of the role of bias for expected predictor error. The blue circles are the "true" conditional probabilities. Overlaid is the "true" decision boundary (the dotted line), the expectations of the fitted values for a linear fit (the sold line), and the expectations of a nonlinear fit (the dashed line). There is ample evidence of bias, but classification accuracy is quite good.

The intent is to work through some of the implications of Equation 4.4. Suppose the true conditional probabilities are known. The dotted line is then the "true" conventional decision boundary located at a conditional probability of .50. Above that boundary, the class of failure is assigned. At or below that boundary, the class of success is assigned. This follows from the requirement of choosing the most probable class. For example, the three blue circles on the far left all imply that the class of failure is assigned. All of the other blue circles, save two near the right boundary, imply that the class of success is assigned.[6]

In what sense is this Bayes classifier optimal? For any blue circle, the assigned class will be correct for some individuals but not others. For example, the far right circle represents a conditional probability of about .40 so that the class assigned is success. But that means that 40% of the time there will be a misclassification for those who fail. For the blue circle immediately to the left, the conditional probability is about .55, so that the class of failure is assigned. But that means that about 45% of the time there is be a classification error for those who succeed.

Under 1-0 loss, one can sum the misclassifications associated with each conditional probability, each weighted by that conditional probability. For the Bayes classifier, the weighted sum will be the smallest sum possible for any set of (nonzero) probabilities weights. Put another way, one arrives at the smallest sum of the losses weighted by the probabilities of misclassification.

In practice, the true conditional probabilities are unknown. We are trying estimates of them. One obstacle is that we are unlikely to have all the important predictors in our dataset. Another obstacle is that we usually do not know the functional forms through which the predictors are related to the response.

A somewhat dated approach is least squares regression applied to the data on hand. Least squares regression is linear in the sense that the predictors are combined as a linear combination to arrive at the fitted values. The functions by which each predictor is related to the response do not have to be linear. But for simplicity and with no important loss of generality, we assume a linear function for the moment. By the same reasoning, we are not considering logistic regression, although with binary outcomes, it is preferable.

The straight heavy line in Figure 4.1 represents the expected values of a least squares fit using the response coded as 1 or 0. The fitted values can serve as an idealized version of a decision boundary from a least squares regression. In this instance, the linear fit captures the true negative relationship. On the average, the probability of failure on parole declines with age.

However, for all but a single conditional probability, the expectations of the fitted values do not overlay the blue circles. For the corresponding ages, estimates of the conditional probabilities are biased. This could result from the linear function being inappropriate and/or the impact of predictors not included. For sixteen of the seventeen ages, estimates of the true conditional probabilities will be systematically too high or too low. If one is trying to understand *how* age may be related to failure on parole, a great deal of information is lost. For example, the increasing risk after

---

[6] The true decision boundary when there are more than two classes is the region where a given outcome has a higher probability than any of the others.

age 30 is completely missed.[7] Clearly, the goals of explanation and understanding are compromised.

How do the biases affect classification accuracy? In fact, expectations of the fitted values do pretty well. Eleven of the seventeen conditional probabilities fall on the same side of both the true decision boundary (i.e., the dotted line) and the corresponding expectations of the fitted values. The ones that do not are shown within the two red ellipses. For example, the two values on the right side are properly assigned the class of parole failure but the regression results would lead to an assigned class of parole success. About 64% of the time, both boundaries classify the same way.[8] Considering the pervasive bias in the least squares results, the classification accuracy may be surprising. The point is that although the actuarial methods used for explanation and for forecasting can rest on the same statistical foundations, the enterprises are quite different. Sometimes, one can be effectively carried out although the other can not.

It is often possible to reduce bias by exploiting nonlinear relationships between the response variable and the predictors. One such relationship is shown by the dashed line in Figure 4.1. On the average, bias is reduced. Using age 30 as an example, the bias associated with the linear fit is the sum of the lengths of the red and green two-headed arrows. The bias associated with the nonlinear fit is the length of the red red arrow alone. One result is that more substantive understanding is gained about how age and recidivism are related. Note that the increase in risk after age 30 is now apparent. There is also an improvement in forecasting accuracy: a little more than 82% of the classifications agree. Another functional form could do even better, and having additional predictors might also help.

4.5.2 Estimation Variance

Expected prediction error involves more than just the impact of bias. There is also the impact of the variance in the fitted values. For a given configuration of predictor values, x_0, the following decomposition can be can be written.

$$\text{EPE}(x_0) = \text{Var}(y_0|x_0) + \text{Var}_T(\hat{y}_0) + \text{Bias}^2(\hat{y}_0) \tag{4.5}$$

There is a lot to unpack. Continuing with the parole failure outcome coded as "1" for fail and "0" for not fail, a 1 or a 0 is the observed value of the response Y. For a given value of age, say, 23 years old, the goal is to estimate the value of the response. Let y_0 be the observed binary response and its corresponding age value be x_0.

Equation 4.5 shows that the expected prediction error is composed of three parts. The first term on the right hand side is the variance of the 1's and 0's around the

---

[7] These might be for individuals disproportionately engaged in domestic violence where the risk at ages beyond 30 can be substantial.

[8] One must be clear that the expectations of the fitted values are being use here. So, the 64% represents an average accuracy.

true conditional probability. 23-year-olds may have, according to $\Pr(G,X)$, a "true" probability of .30 of failing on parole. For 30% of the data realizations for 23-year-olds (or a even a single 23-year-old), a 1 will be observed, and for 70% of the realizations a 0 will be observed. There is variance around the true probability, sometimes called the "irreducible variation" because you are stuck with it.

The second term is the variance in the estimated fitted values over realizations of the training data denoted here by T. This is just the usual variance associated with estimates with one important complication. Under the joint distribution formulation, both the response and the predictors are random variables. One consequence is that the variance in the fitted values tends to be larger than when the predictors are fixed. Another is that when combined with the prospect of nonlinear response functions, the properties of the fitted values are more complicated to address, as we will soon see.

The third term is the square of the bias in the estimate. As such, it is the systematic disparity between the ideal target of estimation and the expected value of the estimate. Researcher often worry most about bias, but in practice either of the other terms in the expression could be substantially larger and, therefore, more worrisome.

4.5.3 The Bias-Variance Tradeoff

The irreducible variance depends on the value of the true probability of failure. That variance shrinks as the probability approaches 1 or 0. The variance of the estimated fitted value depends on properties of the data, the sample size, and the statistical methods employed. For example, in linear regression when predictors have more variability, the sampling variance in the fitted values will be smaller, other things equal. As we saw in Figure 4.1, the square of the bias depends on the "truth," the statistical methods, and the quality of the data.

An important point in practice is that there are ways a researcher can affect the variance and the bias, even for the data already on hand. Consider Figure 4.2, which begins as a reproduction of Figure 3.2. We are back to a scatterplot.

Starting from the left, the three age values that were previously represented by three different indicator variables are now represented by a single indicator variable. For example, if the age of an individual is 16, 17, or 18, the coded value is "1". For all other ages, the coded value is "0". The result is the value of the gray dotted line that is a weighted average of the three gray conditional proportions.[9] The same operations are applied to all subsequent groups of three ages from left to right.

When each age had its own indicator variable, all of the conditional probabilities were estimated in an unbiased fashion for the joint distribution that generated the data. In practice, that distribution will be $\Pr(G,X^*)$ not $\Pr(G,X)$. That is, one can obtain unbiased estimates of the "wrong" conditional probabilities in the sense that the

---

[9] The weights are a function the number of observations for each age.

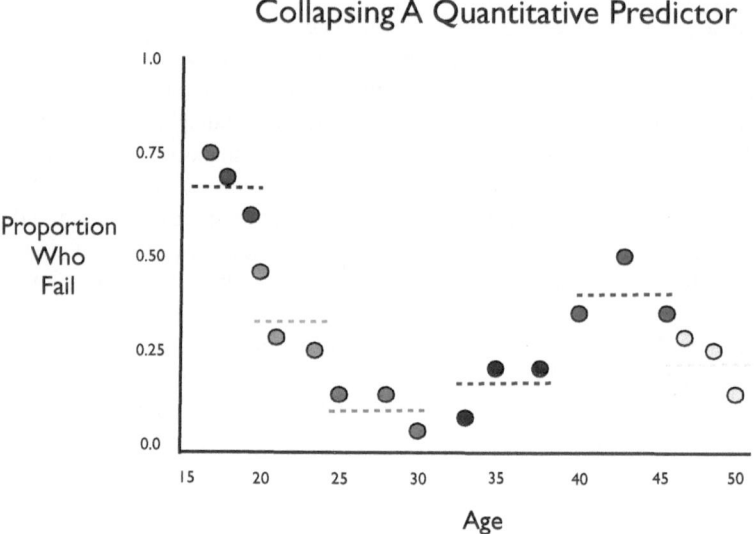

Fig. 4.2 Plotting the proportion who fail by age, collapsing indicator variables. There is more bias but less variance.

data are not a realization from $\Pr(G,X)$. One simply uses the conditional proportions from the data to estimate the conditional probabilities of failure for $\Pr(G,X^*)$.[10]

However, unless all three ages in any of the 3-year age intervals have the same the true proportion who fail, there will be bias associated with each of the collapsed age estimates. At the same time, because each of the dashed line estimates are based on more observations, the estimates will have less sampling variance. They are more stable. Consequently, the expected prediction error could be larger or smaller depending on whether the increase in bias is offset or not by the reduction in the variance. It is entirely possible on the average to have estimates that are closer to their estimation targets when bias is larger, as long as a smaller variance compensates sufficiently. In short, unbiased estimates have no special cachet.

The two ways in which we have constructed indicator variables illustrates the bias-variance tradeoff implicit in Equation 4.5. For a given training dataset, the man-

---

[10] To fix this idea, consider conventional survey sampling. With a simple random sample from a real and finite population, the proportion of 23-year-olds in the sample who failed is an unbiased estimate of the proportion of 23-year-olds in the population who failed. For a finite population, the estimation target is the population conditional proportions. For a joint probability distribution, the estimation target is the conditional probabilities. For the former, the data are called a sample. For the latter, the data are called a realization. But the inferential reasoning is much the same.

ner in which one constructs estimates will often allow one to balance the bias against the variance to reduce the expected prediction error. For example, think back to the earlier discussion of classification trees. With a larger number of terminal nodes, bias can be reduced. But with a larger number of terminal nodes, the observations will be spread more thinly. Many nodes will have far fewer observations. The sampling variance in the vote proportions, and subsequent classes assigned, will be greater. The bias-variance tradeoff figures substantially in forecasting procedures, and more generally can be an important issue in all estimation.

4.5.4 Uncertainty

Equation 4.5 underscores, that as an empirical matter, classification and the forecasts that can follow will vary from realization to realization. In the language of survey sampling, there will be random sampling error. This variation comes from two sources: variance in the estimates, and the irreducible variance. Proper forecasting should include their impact, ideally as forecasting confidence intervals. Decision-makers need to know the "margin of error" in any forecasts they hope to use.

As already noted, the reliance on $\Pr(G, X^*)$ rather $\Pr(G, X)$ requires some reconsideration of what is to be forecasted. The target is no longer the "true" response but an openly-acknowledged approximation. So far, so good.

The overriding problem is that the statistical procedures we will be using compromise conventional forecasting intervals. The bias-variance tradeoff means that bias is likely to be built into any estimates even if the relevant distribution was for G and X. The bias will mean that the usual confidence intervals will not have their stated coverage. The interval will be systematically shifted up or down. A 95% confidence interval, for instance, may cover only 75% of the time.

In addition, we will commonly use the data to help arrive at fitted values that perform well. In effect, we will be engaged in model selection. Model selection can have deleterious effects on statistical inference with often no easy way to make proper adjustments (*Leeb and Pötscher, 2005; 2006, Berk et al, 2010*).

For these and other reasons, one can employ a form of resampling to represent uncertainty. Here is the basic idea:

1. Just as for the nonparametric bootstrap (*Efron and Tibshirani, 1993*), draw with replacement a large number of random samples of size N from the training data.
2. Apply a classifier to each, and save the results.
3. When forecasting, insert the predictor values for each case to be forecasted into each classifier.
4. Count the number of times each case to be forecasted is assigned each class.
5. Forecast the class with the most votes.
6. Report the votes as a way to represent uncertainty (e.g., 52% of the time the class of "fail" was assigned).

These steps simulate what could happen over a large number of realizations of the data. Suppose the outcomes are "fail" or "not" on parole. For one individual, the forecasted class is "fail," assigned 98% of the time over realizations of the data. For another individual, the forecasted class is also "fail," but only 52% of the time over realizations of the data. There is substantially more uncertainty in the forecast for the second individual. The chances of forecasting correctly are only a little better than 50-50. Uncertainty represented in this manner is likely to be understood by decision-makers.

But there is an additional subtlety. The resampling procedure selects entire cases, so that both X and G are random variables. It can be shown that with random X, another source of bias is introduced. It can also be shown that under certain assumptions, the estimates based on $\Pr(G, X^*)$ can be asymptotically unbiased for important features of *that* distribution, but those assumptions are not likely to apply to categorical outcomes.

What one has, therefore, is a way of representing the uncertainty of a *procedure*. The procedure is the classifier. Put another way, we are capturing instability in forecasts resulting from how the data were generated. We are bypassing the matter of bias altogether and focusing exclusively on the variance. Is that acceptable? Once $\Pr(G, X)$ is abandoned, there would seem to be no choice.

In practice, uncertainty could be addressed in two steps. First, when the actuarial methods are being developed, one could use training data to compare how those methods perform when the outcome is known. What is the probability of assigning the correct class? Such information goes to the quality of the method in general. Second, the procedures just outlined can be used to provide information about uncertainty for particular instances when forecasts are needed. Then, the uncertainty can be important information for decision-makers.

4.6 A Bit More on the Joint Probability Model

The centrality $\Pr(G, X)$ and $\Pr(G, X^*)$ may seem somewhat fanciful. But this approach rests on far fewer untestable assumptions than causal models popular in criminal justice research. Moreover, a forecasting enterprise is difficult to formally justify without it. If the intent is to build a forecasting procedure to be applied to subsequent observations, it stands to reason that both the training data and data used in forecasting must be realizations from the same distribution. How can a forecasting procedure built for one distribution be properly used to make forecasts for another distribution?

In practice, things are more complicated. Whether the population from which the training data were drawn is the same as the population for which forecasts are desired is a matter of degree. Recall the earlier discussion about how a variety of factors affecting criminal justice decision-making and outcomes can change over time. Moreover, some parameters of a population subject to change are more important than others.

The parameters for the univariate statistics usually do not matter much. For example, the mean and variance of parolees' ages can differ in the two joint distributions, and the forecasts for the same configurations of predictor values will not differ. But if the associations between the predictors and the response and/or between the predictors change, the forecasts can differ, sometimes dramatically.

There is no way to definitively determine when the distribution responsible for the training data and the distribution responsible for the forecasting data differ in important ways. They are not going to be identical, because with the passage of time, the are inevitable changes in the criminal justice system, as well as variability in the factors that shape criminal behavior. But as discussed in Chapter 2, one can apply some empirical leverage so that quite often, reasonable judgments can be made about whether the two distributions are similar enough for the forecasting task at hand.

A more subtle issue is whether the manner in which the data are realized conforms to the requirement that each observation has a fixed probability of selection. Again, however, there is information that can be brought to bear. For example, a brutal murder committed by an individual on parole may change the way a parole board weighs the risk of releasing individuals on parole. One can think of this as a new population or as new way in which the data are realized. The empirical question, once again, is how much it matters. As before, the primary issue is the relationships among the variables, not their univariate statistics.

The realization process can also be adversely affected when the realizations are no longer independent. For example, perhaps overcrowded prisons push parole boards to become more lenient. With each new denial of parole, the threshold for release is relaxed. This will gradually change the mix of parolees. A more complicated matter is what it does to the formal properties the forecasts. The procedures just described to address uncertainty assume that every case is realized independently of every other case. If the decision made on one parole case alters the probability that the next case will be granted parole, independence is compromised. The way in which uncertainty is characterized is compromised as well.

Unless the dependence is well-understood and built into the bootstrap sampling, there is no good solution. In practice, whatever dependence exists will likely be difficult to document, let alone build into the sampling code. Fortunately, there will often be situations in which one can rule out on substantive grounds important sources of dependence. For example, the statutes and regulations governing parole decisions will often require that decisions be made on a case-by-case basis with public safety the primary concern. The parole board is supposed to judge each case on its merits; it is not supposed to be in the business of regulating the size of prison populations. There should be at least indirect evidence on whether this is effectively true. Sitting in on meetings when parole decisions are made might be a good start. More systematically, one could examine temporal trends in parole decisions from the usual administrative records to help determine if earlier decisions are directly related to later decisions.[11] For what it may be worth, it seems unlikely that such dependence would be substantial.

4.7 Summary

The goal of forecasting requires that one think about the training data, the test data, and the forecasting data as random realizations from the same joint probability distribution. There is no causal model in the usual social science sense, and causality plays no role in the data generation process. One imagines that there can be a limitless number of random realizations of the data and within that framework, estimation, forecasting, and statistical inference can play through.

Practice is far more messy. The training data, test data, and forecasting data may arguably be realizations from the same joint distribution, but key features of that distribution are usually not available to the researcher. The result is that there will likely be bias with respect to the "true" joint probability distribution for any classification or forecasting.

A helpful fallback position is to think about the training data, test data, and forecasting data as realizations from a joint probability distribution whose features *are* available to the researcher. This is not some sleight of hand. Analogies to estimation from survey samples should provide the intuitions needed. For a finite population, a random sample of individuals with different ages can be used to obtain unbiased estimates of, say, median income for each age group, even though income is surely related to more than age.

Yet, there are more complications. Expected prediction error leads naturally to Bayes classifiers, but the estimated conditional probabilities depend on the statistical methods used and the unknown functions linking the response to the predictors. In practice, one must settle for empirically-derived approximations of those conditional probabilities. That too sounds worse than it is because in the end, what one should care most about is the accuracy of the forecasts. That can be directly addressed using test data. As a practical matter, either the forecasts pass muster or they don't.

Properly representing uncertainty raises a number of additional difficulties. One problem is the bias that is virtually inevitable and can even be desirable if a proper balance with the variance is achieved. Another problem is that there does not seem to be any closed form route to proper standard errors. However, resampling procedures have promise and provide at least some protection against reading too much into point estimates.

Figure 4.3 provides a visual summary of the joint probability distribution framework. There is a limitless population having some joint probability distribution $Pr(G, X^*)$, and a function $G = F(X^*)$ associating a response to an available set of

---

[11] For example, one could construct a monthly time series of the proportion of individuals granted parole and examine the autocorrelation in the time series at different lags.

The Joint Probability Distribution Model

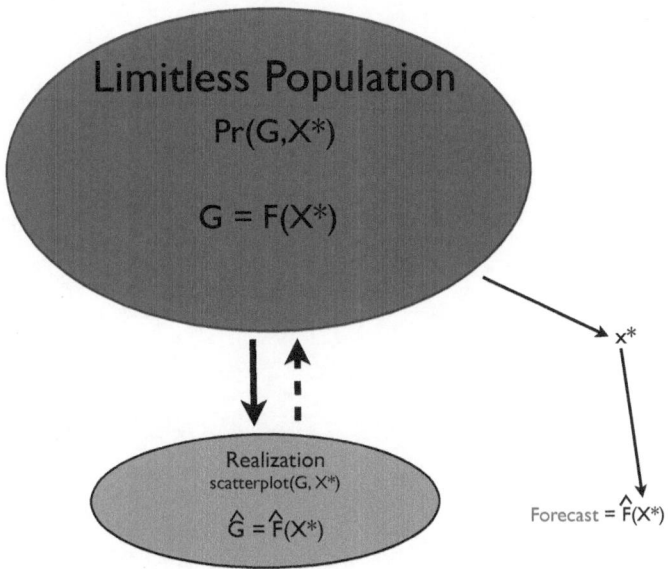

Fig. 4.3 Key elements of joint probability formulation: a limitless population, a random realization of training data and test data, and new predictor values for which a forecast is needed

predictors **X***. The function is "wrong." It is not a representation of some "true" relationship, nor an expression for how the data were generated. It is just a convenient way to approximate the response variable's conditional probabilities.

One can think of the population as what in principle nature could produce. The data are a random realization (shown with the solid arrow) from that population. The data have a multivariate scatterplot corresponding to the joint probability distribution in the population. From the data, one can construct an estimate $\hat{G} = \hat{F}(X^*)$ of how in the population the response is related to the available set predictors. The estimation target is various features of the "wrong" model. Inferences from the realization to the population are represented by the broken arrow. Finally, when some new realization of the predictors is provided, having no known value for the response, the predictor values can be inserted into the estimated function to arrive at a forecast. This is shown on the far right.

In contrast to the usual regression approach in criminal justice forecasting, there is no causal model and no need for one. Consequently there are no disturbances perturbing the response, and no need to obsess over their properties. Uncertainty stems solely from the process by which the response and the predictors are jointly realized from the limitless population.

Chapter 5
Tree-Based Forecasting Methods

Abstract The past two chapters have provided the necessary technical background for a consideration of statistical procedures that can be especially effective in criminal justice forecasting. The joint probability distribution model, data partitioning, and asymmetric costs should now be familiar. These features combine to make tree-based methods the fundamental building blocks for the machine learning procedures discussed. The main focus is random forests. Stochastic gradient boosting and Bayesian trees are discussed briefly as worth competitors to random forests

5.1 Introduction

We will be emphasizing forecasting methods that are "tree-based." This means that classification trees are a key component of the procedures. Classification trees have a long history (*Breiman et al., 1984*), and over the years have proven very effective in identifying complicated associations between a response variable and a set of predictors. However, used by themselves, classification trees can be very unstable, and are usually not a good stand-alone method. Moreover, the recursive nature of the partitioning and implicit reliance on step functions can badly misrepresent additive, smooth relationships such as those commonly assumed with convention linear regression. Still, if the inherent responsiveness of classification trees can be exploited with far greater stability and with a means to better summarize simple smooth functions, one may approach the best of all possible worlds: low bias, low variance, and sensible approximations of key relationships.

5.2 Splitting the Data

We have already discussed recursive partitioning, decision boundaries, and classification trees in a nontechnical manner. It is now time to add some formal details. To begin, how does the procedure decide which partitions to construct?

A good place to start is with the need to subset the data into two groups using the available predictors. The goal is to identify good break points. For a single quantitative predictor with m values, there are $m - 1$ splits that leave the order of the values unchanged. Therefore, $m - 1$ splits on that variable need to be evaluated. For example, if there are 50 distinct ages, there are 49 possible splits that maintain the existing order. The same logic holds for a single ordinal predictor. For a categorical predictor, order does not matter. Consequently, a categorical variable with K categories has $(2^{k-1} - 1)$ possible splits. For example, if there are 5 countries of origin, there are 15 possible splits.

Recursive partitioning begins with all of the training data in the "root node." All possible splits for all available predictors are examined, and the "best" single split over all available predictors is selected. The chosen split is better than the best split of any other predictor, and the data are partitioned accordingly. The same procedure is applied to all subsequent partitions until all observations have been placed in a terminal node. A predictor can be chosen more than once as the partitions are determined. Because the final partitions do not overlap, each case can only be in one terminal node.

What is meant by "best"? The goal is to have as little heterogeneity within a node as possible. The best split, therefore, is the one that reduces heterogeneity the most. Getting to a formal definition requires a few steps.

Using the reasoning in Hastie et al., (*2009:Section 9.2.3*), for any node m, defining partition P_m, having N_m observations, one can estimate

$$\hat{p}_{mk} = \frac{1}{N_m} \sum_{i \in P_m} I(y_i = k), \tag{5.1}$$

which is the proportion of observations in class k in node m, and I denotes an indicator variable equal to "1" when $y_i = k$ and "0" otherwise. One can then classify, as before, by the largest proportion.

But there are a very large number possible partitions implying different proportions for each class. What now? The proportion of observations in class k becomes the argument in a function to characterize the heterogeneity of a node. There are three popular options for this function.

$$\text{Misclassification Error}: \quad \frac{1}{N_m} \sum_{i \in P_m} I(y_i \neq k) \tag{5.2}$$

$$\text{Gini Index}: \quad \sum_{k \neq k'} \hat{p}_{mk} \hat{p}_{mk'} \tag{5.3}$$

$$\text{Cross Entropy or Deviance}: \quad -\sum_{k=1}^{K}\hat{p}_{mk}\log\hat{p}_{mk} \qquad (5.4)$$

Misclassification error is simply the proportion of cases incorrectly classified. Despite its intuitive appeal, it can be insensitive to changes in the terminal proportions because a proportion needs to pass some plurality threshold before the assigned class changes. In the two class case, for instance, a change from .00 to .49 in the proportion who "fail" does not alter the class assigned. It is still "not fail."

The Gini Index or the Cross-Entropy are most commonly used because they are more sensitive to the node proportions. Both take advantage of the arithmetic fact that when the proportions over classes are more alike, their product is larger (e.g., $[.5 \times .5] > [.7 \times .3]$). Intuitively, when the proportions are more alike, there is more heterogeneity. Not surprisingly, the results are usually much the same.

Once all possible splits across all possible variables are evaluated, a decision is made about which split to use. The impact of a split is not just a function of the heterogeneity of a node, however. The importance of each node must also be taken into account. A node in which few cases are likely to fall should be less important than a node in which many cases are likely to fall.

We will define, therefore, the improvement resulting from a split as the heterogeneity of the parent node minus the weighted heterogeneities of two offspring nodes. That is, one can write the benefits of the split, s, for node m as

$$\Delta H(s,m) = H(m) - \Pr(m_{left})H(m_{left}) - \Pr(m_{right})H(m_{right}), \qquad (5.5)$$

where $H(m)$ is the value of the parent heterogeneity for node m; $\Pr(m_{right})$ is the probability of a case falling in the right offspring node; $\Pr(m_{left})$ is the probability of a case falling in the left offspring node; and offspring node heterogeneities are denoted by subscripts $left$ and $right$. The numerical value of Equation 5.5 depends on which measure of heterogeneity is used, but for a given measure of heterogeneity, we want $\Delta H(s,m)$ to be large.

In short, before each partitioning of the data, the value of $\Delta H(s,m)$ is computed for each possible split of each possible predictor. The split and predictor with the largest value are chosen to define the partition. The procedure is applied to the root nodes and all subsequent nodes until there are no other partitions for which $\Delta H(s,m)$ is large enough to matter or until $\Delta H(s,m)$ can no longer be computed because all of the terminal nodes each have only one observation.

If we intended to use classification trees as our classification and forecasting tool, there are additional details to consider. For example, how would terminal nodes with very few observations be handled? One might "prune" such nodes. But classification trees are just the beginning; There are many other issues to consider.

5.3 Building the Costs of Classification Errors

Recall that some classification errors are more costly than others, and that it can be critical to build in the relative costs of such errors at the very beginning when a forecasting procedure is being developed. This is easy to do to in many popular implementations of classifications trees.

We begin by defining in more detail a $K \times K$ loss matrix $\mathbf{W}$.[1] Suppose for the moment that $K = 2$. The responses outcomes are "fail" or "not". When the forecasting procedure misses a failure, one has a false negative. When the forecasting procedure incorrectly flags a failure, one has a false positive. Then $\mathbf{W}$ is

$$\begin{bmatrix} 0 & R_{fn} \\ R_{fp} & 0 \end{bmatrix}$$

where the entries along the main diagonal zero, and the off-diagonal elements contain costs for false positives (i.e., R_{fp}) and false negatives (i.e., R_{fn}). The units do not matter. What matters is the ratio of the two. For example, R_{fn} could be 10, and R_{fp} could be 1. False negatives are 10 times more costly than false positives. Put another way, 10 false positives have the same cost as 1 false negative.

Such reasoning can be easily extended when there are more than two classes. But all two-way comparisons need to be addressed. Below is the case when $K = 3$. Numerical subscripts for rows and columns are used because it is no longer clear what a false positive or false negative is. There would need to be a numerical value for each off-diagonal element so that the ratio between any pair of relative costs is represented. For example, R_{23} could be 2 and R_{21} could be 3 for a cost ratio of 3 to 2.

$$\begin{bmatrix} 0 & R_{12} & R_{13} \\ R_{21} & 0 & R_{23} \\ R_{31} & R_{32} & 0 \end{bmatrix}$$

There are several ways in which the information in a loss matrix can be introduced. The easiest way is to use the off-diagonal elements in $\mathbf{W}$ to weight the probabilities of the Bayes classifier. However, that only can affect the classes assigned and not the full tree-building process. Interpretations of the tree structure would likely be based on the wrong tree. It is better to build in the relative losses at the very beginning.

There are two such approaches that in practice can lead to the same results. One is to weight the data by the relative costs. The other is to alter directly the prior distribution of the response to reflect those costs. For example, suppose 30% of the cases in the training data "fail" and 75% do not. But when time comes for the analysis, the 30-70 prior could be changed to 45-55 so that failures are made relatively more numerous. Altering the prior implies reweighting. For both approaches, conse-

---

[1] In some treatments (*Hastie, et al, 2009: 310-311*), the matrix is denoted by $\mathbf{L}$ despite the same symbol being used for the loss function in discussions of expected prediction error.

quently, observations with more costly classification errors are made relatively more important. Which method one uses will often depend on how the classification tree software is structured.

Because in our treatment, a single classification tree by itself is not the basis for classification and forecasting, we will move on. Suffice it to say that there is a third approach, discussed later, better suited for the perspective we will take. The basic idea is to sample the data so that the prior distribution of the response is changed.

5.4 Classification Tables

For our purposes, the most important output from a classification tree compares the classes assigned to the actual classes. There are the two settings. For the training data, there can be classes assigned and classes observed. For forecasting data, there can be classes forecasted and then, at some later time, classes observed. We will emphasize the latter.

A cross-tabulation of the forecasted class and the subsequently-observed class can be usefully presented as a contingency table, sometimes called a classification table or a confusion table. Table 5.1 is an illustration for a binary response variable. The two response classes are "success" and "failure." The letters a through d are counts of the number of cases. For example, a is the number of cases for which a failure was forecasted and a failure was observed.

	Forecasted Failure	Forecasted Success	Model Error
Observed Failure	a	b	$b/(a+b)$
Observed Success	c	d	$c/(c+d)$
Use Error	$c/(a+c)$	$b/(b+d)$	Overall Error $= \frac{(b+c)}{(a+b+c+d)}$

Table 5.1 A classification table for a binary response variable. The letters represent counts of the number of observations. There are two outcomes: "success" and "failure."

In the bottom right cell is the proportion of cases for which the forecasted class and the observed class is not the same. This is the overall forecasting error. It treats all forecasting errors the same. Typically they are not. Hence, there is usually more interest among decision-makers in the calculations along the margins of the table.

Under "Model Error" are the conditional proportions for the actual failures and actual successes, respectively, whose forecasted class is not the same as the observed class. Given an observed failure, what fraction of the time does the forecasting procedure forecast a success? Given an observed success, what fraction of the time does the forecasting procedure forecast a failure? Both are measures of the performance of the forecasting procedure given the truth, and both proportions should be small. Ideally, the forecasting procedure should be able to identify groups with high concentrations of either success or failures respectively. Model error is often the focus when a forecasting procedure is being "validated," although "evaluated" is a far less loaded term.

Under "Use Error," one conditions on the forecast, not the truth. Given a forecast of failure, what fraction of the time does the procedure get it wrong? Given a forecast of success, what fraction of the time does the procedure get it wrong? These error rates help inform decision-makers about how the forecasting procedure will perform in practice. How credible is a forecast of success? How credible is a forecast of failure? Use Error is commonly the focus in the field.

Why is it necessary to unpack overall forecasting error in these ways? Overall forecasting error equates the costs of false positives with the costs of false negatives. Typically they are not the same, which means that one should condition on the actual outcome and/or the forecasted outcome when calculating an error rate. For example, forecasts of success may be wrong 10% of the time, while forecasts of failure may be wrong 25% of the time. Likewise, the procedure may fail to correctly identify failures 45% of the time, and fail to correctly identify success 15% of the time. We will later see how such figures play out in some real world examples.

5.5 Ensembles of Trees: Random Forests

Recall two important limitations of classification trees: instability, and reliance on step functions when the more appropriate functions are smooth. Both can be usefully addressed with a large number of classification trees. We consider the instability first.

Resampling from training data was mentioned earlier as a useful tool. For random forests, a large number of samples, with the same number of observations as the training sample, are drawn with replacement. The samples are nearly independent. On the average, about a third of the observations are not selected each time. A classification tree is grown from each sample and then the class assigned to each observation is determined by a vote taken *over the trees*. By working with a vote over a large number of trees, random sampling errors tend to cancel out. The process of averaging some summary statistic over a large number of random samples of the training data is called "bagging" (*Breiman, 1996*).

Figure 5.1 addresses the difficulties that can be caused by step functions. The response is failure on probation. The predictor is age. The dashed line is some S-shaped function that we wish to approximate. Suppose a classification tree splits the data at age 37. The green lines shows, with a bit of artistic license, the step function. (The horizontal segment below 0 should fall right on top of the graph's horizontal axis. But then it would not be visible.) One can see that the step function is a poor approximation of the S-shaped function.

The figure shows, in different colors, three other step functions from three other random samples. The break points happen to differ. The same artistic license is used. The four step functions *as a group* do a much better job than any single step

Step Function Approximation

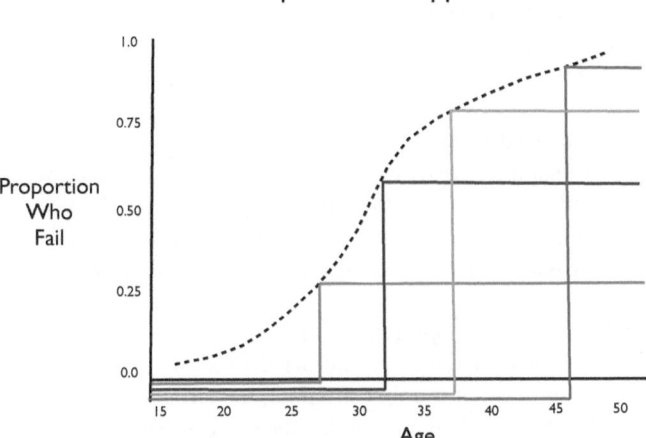

Fig. 5.1 Using four step functions to approximation a nonlinear response to age. The approximation will generally improve as the number of step functions increases.

function approximating the S-shaped function. A 100 steps functions would do still better. And that's the point.

But there are a lot more going on. Here is a synopsis of random forest algorithm.

1. There is a training dataset with N observations. A random sample of size N is drawn with replacement from the training data. The selected observations are used to grow a classification tree. Observations not selected are saved as the "out-of-bag" (OOB) data. These can be used as test data for that tree and will on the average be about a third the size of the original training data.
2. A random sample of predictors is drawn. The sample is often very small (e.g., 3 predictors).
3. The first partition of a classification tree is grown, selecting the best split from one of the random subset of predictors. The result is the two subsets of the data that maximize the reduction in the Gini index.
4. Steps 2 and 3 are repeated for all subsequent partitions until further partitions do not improve the model's fit.
5. The Bayes classifier is used with each terminal node to assign a class.
6. The OOB data are "dropped" down the tree, and each observation is assigned the class associated with the terminal node in which that observation lands. The result is the predicted class for each observation in the OOB data for a given tree.
7. Steps 1 through 6 are repeated a large number of times to produce a large number of classification trees.

8. For each observation, classification is by vote over all trees when that observation is OOB. The class with the most votes is chosen.

By building a forest from a large number of classification trees, many of the assets of recursive partitioning carry over. For example, strongly nonlinear functions and high order interaction effects can be examined with no need to specify either in advance.

But new virtues are also introduced. The large number of trees derived from a large number of random samples provides opportunities for otherwise overlooked relationships to be found. Associations that might appear to be weak in a given sample because of random sampling error may surface more importantly in other samples. Each sample provides another, and somewhat different, look at features of the population.

By sampling predictors at each split, random forests allows a wide variety of trees to be grown (*Ho, 1998*). Stronger predictors do not necessarily mask weak predictors because for a substantial number of splits, the stronger variables are not among those randomly selected. This can make for more flexible model building especially when the training samples have a large number of observations.

Random forests allows the relative costs of forecasting errors to be built directly into the algorithm in one of several ways. Craft lore suggests that the best way is by using a stratified sampling approach each time the training data are sampled with replacement. By over-sampling some response classes and under-sampling others, one can alter the prior distribution of the response and reweight the data. More detail will be provided later through some examples.

As the number of trees increases without limit, random forests does not overfit (*Breiman, 2001a: 7*). One key reason is the OOB data. For any given tree, the data used to grow the tree are not "resubstituted" for purposes of classification. Classification is undertaken with observations precluded at random from the tree growing process.

Finally, the OOB data can also be a very good approximation of true test data. Consequently, classification/confusion tables from random forecasts show a good approximation of the performance with real forecasts, not how well the procedure reproduces the training data. The result is a more honest assessment that will be especially useful when conventional test data are unavailable.

5.5.1 Variable Importance

Especially when used for forecasting, the most important output from classifiers is the classes assigned to different observations. But for criminal justice decision makers, the forecasts alone may not be sufficient. Because of legal and administrative concerns, knowing the importance for forecasting of each predictor can be very instructive. How important, for instance, is gender? What about age or prior record?

A useful operationalization of a predictor's importance is its contribution to forecasting accuracy. For random forecasts, this is addressed in a clever manner. Forecasting accuracy is first computed using all of the available predictors. Then, one at a time each predictor is doctored so that it cannot contribute to foresting accuracy. The resulting drop in forecasting accuracy for each predictor is a measure of importance. Here is the algorithm.

1. For a given tree, compute as usual the assigned class for each case in the OOB data.
2. Compute the proportions of cases misclassified for each response class. These proportions serve as a forecasting accuracy baseline when the full set of the predictors are used to grow the tree and contribute to forecasting accuracy.
3. Randomly permute the values of a given predictor over all observations. This will on the average make that predictor unrelated to the response variable.
4. Making no other changes in the tree, compute again the assigned class for each case in the OOB data. Many of the classes assigned will likely differ from the baseline assignments because the shuffled predictor no longer can help determine the terminal nodes in which observations fall.
5. Compute the proportions of cases misclassified for each response class. These proportions serve as measures of forecasting accuracy when the given predictor is randomly shuffled.
6. Compute the increase in the proportion misclassified compared to when all predictors are used as a measure of that variable's importance for each response class.
7. Repeat from Step 3 for each predictor.
8. Repeat from Step 1 for each tree in the random forest.
9. For each predictor, compute average decline in forecasting accuracy over trees.[2]

Output from the importance calculations can provided in several different formats. First, the output can be displayed separately for each response class. This has the significant advantage of not mixing outcomes when they can differ substantially in forecasting-error costs. In Chapter 2, Figure 2.1 was such as display. Predictors were ranked by the reduction in forecasting accuracy for the response class of "violent crime" when each predictor was in turn randomly shuffled. Recall that the sum of the reductions was less than the total contribution of the full set of predictors because when predictors are correlated, there is shared forecasting accuracy that cannot be attached uniquely to any single predictor. Note also that the ensemble of trees did not change with the shuffling. Only the predictor values provided to the existing ensemble changed.[3]

---

[2] In R, the code for a variable importance plot would take on something like the following form: *varImpPlot(rf1, type=1, class="Fail", scale=F, main="Importance Plot for Violent Outcome")*, where *rf1* is the name of the saved output object from random forests, *type=1* calls for the mean decrease in accuracy, *class="Fail"* is the outcome class *"Fail"* for which the importance measures are being requested, *Scale=F* requests no standardizations, and *main="Importance Plot for Violent Outcome"* is the plot's heading.

[3] It is sometimes desirable in regression analysis to drop one or more regressors from a model to see how a smaller model performs compared to the larger model. When this is done, both the set of predictors and the model itself change. The impact of the predictors and the model are

Second, some researchers prefer to work with a standardized version of variable importance. The tree-by-tree reductions in accuracy used in the Step 9 averaging can also be used to compute the standard deviation of the tree-by-tree reductions. Dividing the accuracy reductions by this standard deviation leads to a standardized average reduction.

Without standardization, predictors with a larger number of distinct values tend to be evaluated as more important. One can argue whether this matters in practice. If the intent is to represent the forecasting importance of different predictors for the data on hand, standardization seems misleading. If the intent is to make more general claims about the forecasting importance of different predictors, standardization seems appropriate. Because the applications emphasized here are dataset-specific, the non-standardized formulation is probably preferable.

Third, one can obtain in standardized or non standardized form the reduction in forecasting accuracy averaged over the full set of response classes (e.g., violent crime, nonviolent crime, no crime). Outcomes with potentially very different costs for incorrect forecasts are combined as if the costs were the same. Consequently, this option is typically undesirable.

Fourth, one may decide that instead of defining importance by the contribution to forecasting accuracy, contribution to fit should be used. The output then can be in units of the Gini Index. One can learn how much the Gini Index increases when in turn each predictor is shuffled. If the goal of a random forests analysis is forecasting, this option seems to be wide of the mark.

5.5.2 Response Functions

A second supplemental output from random forests is how each predictor in turn is related to the response, all other predictors held constant. In effect, the relevant algorithm manipulates the value of each predictor in turn, but nothing else, and records what happens to the average response. Here are the steps.

1. For a given predictor with M values, construct M special datasets, setting the predictor values to each value m in turn and fixing the other predictors at their existing values. For example, if the predictor is years of age, M might be 25, and there would be 25 datasets, one for each year of the 25 age values. In each dataset, age would be set to one of the 25 age values for all observations (e.g., 18 years old), whether that age were true or not. The rest of the predictors would be fixed at their existing values.
2. Using a constructed dataset with a given m (e.g., 22 years of age), and the random forest output, compute the assigned class for each observation.
3. Compute the proportion p_k, which is the proportion of observations in class k, for each of the K classes.

---

confounded. When in random forests predictors are shuffled before being dropped down a fixed ensemble of trees, there is no such confounding.

4. Compute

$$f_k(m) = \log p_k(m) - \frac{1}{K} \sum_{l=1}^{K} \log p_l(m), \qquad (5.6)$$

where the function in log units is the difference between the proportion computed for value m of a given predictor and the average proportion over the K classes for that predictor (*Hastie et al., 2009: 370*). There will be one such function for each response class k, each in logits centered on the average logit.[4] This is analogous to how in analysis of variance the effects of different levels of a treatment are defined; their effects sum to zero. It is sometimes helpful to solve for p_k for different values of m to see how the probability itself changes. The *change* in the probability is a valid measure, but because of the centering, the value of the probability is not.

5. Repeat Steps 2 through 4 for each of the M values.
6. For each response class k, plot the logits from Step 4 for each m against the M values of the predictor.
7. Repeat Steps 1 through 6 for each predictor.

Figure 2.2 in Chapter 2 was an example for the predictor IQ. Recall that the response function was substantially nonlinear — negative up to IQ values of around 100 and flat thereafter. The shape of the function was not anticipated by existing subject-matter theory and may been a surprise to some readers.[5]

5.5.3 Forecasting

With a random forest in hand, forecasting is relatively straightforward. There is a set of observations one might call forecasting data. Predictor values from the forecasting data are provided to each tree in the forest. Every observation is placed in a terminal node according to its predictor values and assigned the class of the terminal node previously determined when the tree was grown. The class forecasted is determined by a vote over trees.

For example, if there are 500 trees, a given observation will be placed in 500 terminal nodes and assigned the class of each. The class forecasted is determined by a vote over all the trees. The class with the most votes is the class forecasted. Output can include not just the forecasted class, but the votes over the 500 trees for each class. The votes provide information on forecasting uncertainty, to which we now turn.

---

[4] This avoids the problem of having to choose a reference category.

[5] In R, the code would look something like this: *partialPlot(rf1, pred.data=temp2, x.var=Age, which.class="Fail", main="Dependence Plot for Age")*, where *rf1* is the name of the saved random forests output object, *pred.data=temp2* calls the input dataset, usually the training data (here, *temp2*), *x.var=Age* indicates that the plot is being requested for the predictor called "Age", *which.class="Fail"* specifies the outcome class, here, "Fail", for which the plot is being requested, and *main="Dependence Plot for Age"* is the heading of the plot.

5.5.4 Procedure Uncertainty

Broadly viewed, the forecasting being discussed is subject to two forms of uncertainty. The first is *procedure* uncertainty, which plays a key role as the forecasting procedure is being developed. What are the chances that the procedure will correctly identify a given outcome class, such as committing a violent crime? In addition, what are the chances that when a forecast is made, it is correct? For example, if certain individuals are forecasted to be crime-free, what are the chances that the forecast is correct? Both kinds of assessments require that the outcome being forecasted is known. Test data or OOB data are essential.

When a random forest is grown, the confusion table provides information on procedure uncertainty. The table is constructed from the OOB data, so the counts can represent real forecasting. Because for the OOB data the outcomes are known, procedure uncertainty can be directly ascertained. One might learn, for instance, that individuals who commit violent crime are correctly identified 60% of the time; that individuals who commit nonviolent crimes are correctly identified 75% of the time; and individuals who commit no crime are correctly identified 85% of the time. But how good is that?

A baseline for purposes of comparison is the marginal distribution of the response. For example, suppose 10% of the outcomes are violent crimes. Violent crimes are quite rare and individuals who commit those crimes are probably hard to identify *a priori*. (If they were easy to identify in advance, there would be no need to develop a forecasting procedure.) If random forests can find such individuals with 60% accuracy, that may be impressive.

There will be some applications where the performance is not impressive. But predictors are chosen in part because they are thought to be related to the outcome; there are likely to be gains relative to a baseline. Performance is most likely to be striking for the outcomes given more weight in the analysis. At the same time, there will sometimes be instances in which excellent performance for the outcomes given more weight is coupled with modest performance for outcomes given less weight, especially if the former are rare relative to the latter.

The comparisons just examined only address how the model performs when the outcome is known. How likely is it that the forecasting procedure will correctly identify a case with a particular outcome? Decision-makers also want to know how well the forecasting procedure actually forecasts. That is, when a particular outcome is forecasted, how likely is that forecast to be correct?

For example, one might find that when a violent crime is forecasted, the forecast is correct 50% of the time. When a nonviolent crime is forecasted, the forecast is correct 60% of the time. When no crime is forecasted, the forecast is correct 80% of the time. These percentages are computed by conditioning on the *forecast*, not the outcome.

The marginal distribution can again be used, but in a somewhat different fashion to provide baselines. If, ignoring the predictors, a forecast of violent crime is made, it will be right, say, about 10% of the time. If, ignoring the predictors, a forecast of nonviolent crime is made, it will be right, say, 40% of the time. If, ignoring the predictors, a forecast of no crime is made, it will be right, say, 50% of the time. In this hypothetical, the forecasts from random forests are much more accurate than these baseline forecasts — 50% correct, 60% correct, and 80% correct respectively. The improvement for forecasts of violence is especially impressive.

In short, there are two different questions that are sometimes confused. The first question asks how well the forecasting procedure reproduces the known outcomes. The second question is how accurate the forecasts turn out to be. The first can provide useful information as the forecasting model is being developed. The second can provide useful information about how the forecasting model will performs in use.

5.5.5 Forecast Uncertainty

The second form of uncertainty is *forecast* uncertainty; there is uncertainty when the forecasts for *given* cases are used to inform decisions about those cases. The forecasting procedures should have already been deemed sufficiently accurate. Nevertheless, any forecast could be wrong. It is important, therefore, to have some indication of the likelihood of forecasting error as each forecast is made. There are no test data because the outcome is not known. Yet, random forests provides important information on this matter.

Recall that we envision the training data as a random realization from a joint distribution of G and X^*. When the training data are randomly sampled by the random forests algorithm, one has a simulated version of this realization process. There is an instructive correspondence between the underlying thought experiment and the machine learning algorithm.

With each simulated realization of the training data, the tree constructed will almost certainly differ, at least a bit. There is then the prospect of the assigned classes differing as well. For one classification tree, males 30 years of age might be classified as violent, whereas another classification tree might classify them as nonviolent. It follows that vote, over trees, to finally determine the assigned classes, incorporates this chance variability.

Additional uncertainty in forecasts is introduced when the random forest algorithm randomly samples predictors. For any given sample of the training data, the tree that is grown can vary with the predictors randomly available as each data partition is determined. Frankly, there is no apparent correspondence between random samples of predictors and the underlying throught experiment usually applied to data realizations. Nevertheless, because such sampling is built into the algorithm, it should be taken into account.

There are no expressions from which one can compute proper standard errors for random forests forecasts and no way, therefore, to directly construct conventional error bands for the forecasts. But the resampling of training data and predictors can provide the basis for an instructive alternative.

Consider the forecast for a single individual. Suppose that 95% of the votes over trees are for a single class. One can infer that only 5% of the time will the random forests trees forecast another class. Suppose 60% of the votes over the trees are for that class. Now, 40% of the time the random forests trees will forecast a different class. There is substantially more uncertainty. Suppose that 10% of the votes over trees are for that class. Then, one can infer that 90% of the time random forests will forecast a different class, and one would do considerably better forecasting some other class.

It follows that the votes provide information on the variability of a forecast that results from the random forests *procedure*. Uncertainty comes from random realizations of the data and from random realization of the predictors included. It can be useful, therefore, to construct a table or histogram of the vote percentages over cases to get an broad sense of how credible the forecasted classes are likely to be. We will examine some examples later, but highly-skewed distributions are common. It can be difficult to get lopsided votes in favor of uncommon classes (e.g., arrested for homicide).

Probably more directly relevant for decision-makers is the vote for *given* cases. If each outcome class gets the same proportion of votes over trees, the uncertainty is as large as it can be, and there is no guidance on which class to forecast. Only if one class receives a substantial plurality is random forecasts providing useful information for decisions about a given case; uncertainty is low. Put yet another way, unless one class receives a substantial plurality, the forecast obtained could easily have been different.

One must be clear that estimation bias plays no role, and there are no conventional error bands nor talk about coverage for the "true" forecast. This is not a matter of getting the forecast right. The votes provide information on the *stability* of the forecasts over trees for a given case. Again, it is a statement about how the random forests *procedure* performs for given forecasts.

In practice, forecasts of future behavior will be one of several factors taken into account when decisions are made. There will typically be no formal way to weight the factors. This is where forecast stability can helpful. For example, a sentencing judge may wish take very seriously a defendant's cooperation with police as a mitigator for the sentence given. But, how should that be balanced against a forecast of future dangerousness? If the forecast could easily have been different in the sense just discussed, the relative importance of the cooperation should be increased substantially. If the forecast is effectively the only forecast made by a very large number of trees, the relative importance of the forecast should be increased substantially.

For a defendant's cooperation with the police, the issue is sentence mitigation. There is no intent revise the forecast of future dangerousness. If the intent is to revise the actuarial forecast, the revised forecast should be based on information that is not already incorporated into the forecasts or not already discarded as unimportant by the actuarial methods. As noted earlier, even very simple statistical procedures will on average forecast more accurately than informal, even clinical, procedures. It is, therefore, counterproductive to alter actuarial results unless there is new and compelling information not considered when the actuarial forecasts were made. A show of genuine contrition may be an illustration. In short, one should not second guess the forecasts unless there is very important information that was not taken into account when the forecast was made.

5.6 Tree-Based Alternatives to Random Forests

There is an ongoing debate about which classifiers are the most accurate. Truth be told, it is hard to tell. Compelling theoretical results one way or another do not exist, and in classification/forecasting competitions, performace is heavily dependent on particular features of the datasets used. Winners in one competition can be losers in the next. And in the end, performance differences among the better procedures are usually small in practical terms (e.g., *Culp et al., 2007*).

Random forests is a powerful classifier, relatively easy to use, whose requirements, capabilities and output connect especially well with needs of criminal justice decision-makers. That is why random forests is the machine learning method-of-choice here. The ability to easily accommodate asymmetric costs is one key example.

But there are other very attractive classifiers within machine learning traditions. Some compete well against random forests and have many of the same capabilities. With future development, one or more of these could become the criminal justice method-of-choice. Anticipating this possibility, we turn briefly to two of the most promising tree-based competitors. Other competitors that do not build on tree ensembles are not discussed. Space constraints preclude an accessible exposition, and there are excellent treatments elsewhere (Hastie et al., 2009).

5.6.1 Stochastic Gradient Boosting

Random forests strives to construct its many classification trees independently of one another, and for all practical purposes succeeds. Then, classes are assigned to cases by a simple vote over trees. All trees are implicitly given the same weight in the voting; one tree, one vote.

A powerful alternative, called "boosting," combines trees linked in a sequential fashion. The links are provided by each tree's residuals — the differences between a tree's fitted values and the observed responses. Trees are grown one after another such that the residuals from one tree become the "working response" for the next.

Larger residuals imply a less satisfactory fit. For any given tree, therefore, cases that are more difficult to accurately classify are given more weight. Then, the fitted values from each tree are used to revise the previous set of fitted values. In the end, the fitted values are a linear combination of the sequence of fitted values with weights to maximize classification accuracy. Fitted values that improve performance more are given more weight.

There are many different kinds of boosting, even within the subset that works with ensembles of trees. They can vary in a variety ways, implying complicated tradeoffs that are likely to be highly data dependent. We will use stochastic gradient boosting (*Friedman, 2002*) to represent these methods because on balance, it is worthy competitor to random forests. There is also several good software choices available even within R, although here too there can be important tradeoffs.[6]

Consider a training dataset with N observations and p variables, including the response $\mathbf{Y}$ and the predictors $\mathbf{X}^*$. The response is binary. Suppose "fail" is coded as "1" and "succeed" is coded as "0." Here is an outline of the procedure.

1. Initialize the procedure with sensible fitted values before the analysis begins. One option is to use the overall proportion of cases that fail as the initial fitted value for each case.
2. Randomly select without replacement n cases from the dataset, where n is less than the total number of observations N. Note that this is a simple random sample (i.e., sampling without replacement), not a sample with replacement. Often about half of the observations are selected.[7]
3. Compute the negative gradient, also called "pseudo residuals," by subtracting each fitted value from its corresponding observed value of 1 or 0. The residual will be quantitive not categorical.
4. Using the randomly-selected observations, fit a regression tree to the pseudo residuals.
5. Use the mean in each terminal node who fail as an estimate of the probability of failure.
6. Still using the sampled data, update the fitted values as the sum of the existing fitted values and the new fitted value with the latter weighted to get the best fit.
7. Repeat the process steps 2 through 6 until the fitted values no longer improve a meaningful amount.
8. Use the fitted probability estimates as is, or apply the Bayes classifier to get the assigned class.

Unlike conventional iterative procedures in statistics, stochastic gradient boosting does not converge. But there are empirical ways to determine when additional passes through the data are not likely to improve the fit. As an empirical matter, the lack of convergence does not seem to be a problem, but misleading results can follow if a substantial number of additional trees are grown despite no demonstrable improvement in performance (*Mease et al., 2007*).

Stochastic gradient boosting for categorical responses is a powerful machine learning procedure that has many interesting properties. At the moment, however, there does not seem to be easily accessible implementations that allow for more than two response categories and/or principled ways to introduce asymmetric costs

---

[6] The name "boosting" comes from ability of the procedure to "boost" the performance of otherwise "weak learners" in the process by which they are aggregated.

[7] This serves much the same purpose as the sampling with replacement used in random forests. A smaller sample is adequate because when sampling without replacement no case is selected more than once; there are no "duplicates."

(although this could change at any time). Both are requirements for the kinds of criminal justice forecasts emphasized here. For example, *ada*, *gbm* and *mboot* are excellent boosting procedures available in R, but currently are not fully suitable for our purposes (*Culp et al., 2006; Ridgeway, 2007; Hothorn et al., 2011*).

5.6.2 Bayesian Trees

Random forests clearly falls within frequentist statistical traditions. The data are random realizations composed of random variables, and the parameters of the joint distribution $\Pr(G, X^*)$ are treated as fixed. There is uncertainty in the forecasts solely because the realized data could have been different by chance.

Consistent with Bayesian traditions, Bayesian trees turns this on its head. The data on hand are treated as fixed. They are not realizations from anything and whether they could have been different is irrelevant. What you see is what you get. It is the parameters of $\Pr(G, X^*)$ that are random and, therefore, the source of uncertainty. Why they are random is beyond to scope of this brief discussion, but as rough approximation they are random because each parameter is seen not as a having a single value, but a distribution of values. One account makes each distribution a representation of a researcher's beliefs about the probabilities of different possible parameter values.

As with random forests, a large number of classification trees are grown. But the trees differ not because the training data are sampled, but because chance is introduced into the tree-growing process in several ways. The issues can be quite technical and are beyond the scope of our discussion (*Chipman et al., 1998; 2010*). But here is the general idea.

Consider a single classification tree as discussed above. Beginning with all of the observations in the "root node," an initial decision is whether to partition the data into two subsets. If the data are partitioned, the same decision needs to be made for the two offspring nodes. If either of those is partitioned, a new splitting decision needs to be made. Such decisions are made until all of the terminal nodes are determined.

With the Bayesian approach, whether to partition the data in any node is decided, in effect, by a flip of a special kind of coin whose probability of coming up "split" declines as the tree grows larger. The intent is to make larger trees less likely than smaller trees. Smaller trees will usually have more bias but less variance, which given the substantial instability of classification trees, is usually a sensible tradeoff.

Once a decision has been made to partition the data in a node, the nature of that split must be determined. In the spirit of random forests, one predictor is chosen at random as well as the value at which the split will be made. For example, if age is the predictor chosen, the partitioning might be at a value of 30. Cases with ages less than 30 might be placed in one partition, and ages 30 or older might be placed in the other. If the predictor chosen is race/ethnicity, Blacks might be placed in one partition, and Anglos, Hispanics, and Asians might be placed in the other.

The probability that a particular split will be made can take into account the number of distinct values for each possible predictor. Predictors with more distinct values can be assigned lower probabilities so that all possible splits across all predictors are equally likely. That is, unless this weighting is undertaken, predictors with a greater number of values are more likely to be chosen as the splitting variable. Other factors can be taken into account, such as the desire to keep the number of predictors used by a given tree small.

Of key interest is the response variable proportion computed within each terminal node. For our purposes, these are the conditional probabilities for one of the two outcomes, such as failure on parole. A probability distribution is also imposed here, often a beta distribution. An alternative is a uniform distribution. For each terminal node, the response variable proportion is drawn from one such distribution.

With both the probability of a split, and the probability of a particular split, one can in principle generate all possible trees. In practice, computational constraints intervene so that a very large, but not exhaustive, number of trees is generated via simulation. This permits estimates of the probability that any given tree will be selected. These estimates can be used to construct a prior distribution of possible trees with some trees more likely than others. Then, the prior helps to prevent rare and hence, unimportant trees from substantially affecting the overall results.

In short, uncertainty is introduced in four ways: (1) through the probability of any split, (2) through the probability of a particular split, (3) through the probability of a particular proportion in each terminal node, and (4) through the prior probability for each possible tree. The four probabilities are then used to construct an ensemble of trees — a Bayesian forest.

There is little that can be directly learned from the Bayesian forests. Just as in random forests, the information provided by each tree needs to be compiled in an accessible form. One could once again employ some kind of voting procedure. Voting can be seen as an additive process in which for each case, the votes over trees for each possible class are summed. Bayesian trees employs an addition process but instead of combining votes, combines the proportions in the terminal nodes. For example, if the proportion of "failures" in a terminal node is .65, the value of .65 is used directly rather than as a single vote in favor of the response class of "failure." The additive procedure is called "backfitting." Here in very broad terms is what is done.

It all begins with the construction of a new and special dataset. Suppose there are 500 observations in the training data, and 50 Bayesian trees are grown. For *each* of the 50 trees, there are 500 estimates of the binary response variable proportion depending on the terminal node in which the observational falls. One can think of the set of 500 fitted values for a given tree as a new kind *predictor* variable. Over the 50 trees, there are 50 such predictor variables. These can be assembled as a new predictor matrix $\mathbf{Z}$ with 500 rows and 50 columns. The response variable is, as before, the observed binary outcome $\mathbf{G}$ over the 500 observations.

One can then, in effect, regress the binary outcome on the 50 predictors with, for instance, a probit link. Backfitting is used rather than more conventional numerical methods because the functional forms by which the predictors are each related to the response are determined empirically; they are not specified in advance as in parametric regression.

The backfitting algorithm cycles through each predictor one at a time employing a bivariate smoother until further iterations no longer improve the fit. For each smoother application, the response is very much like the pseudo residuals used in boosting. They are what is "left over" in the response after systematic relationships from earlier estimates are subtracted off. In this example, there might be 225 iterations before convergence. All predictors might cycle through 4 times, and 25 predictors might cycle through 5 times. A function of the fitted values, such as the fitted proportions, is the basis of any forecasts.[8]

Bayesian forests can be grown with existing software in R (i.e., with *bart*, which stands for Bayesian Additive Regression Trees). It runs well, and for binary outcomes performs at about the same level of accuracy as random forests and stochastic gradient boosting. One of its major strengths is that the Bayesian framework leads directly and naturally to conventional Bayesian statistical inference. Statistical inference can be a problem for random forests and stochastic gradient boosting. At this point, the major drawback of Bayesian forests are no allowance is made for asymmetric costs or for more than two response categories.[9]

5.7 Why Do Ensembles of Classification Trees Work So Well?

In a wide variety of social science applications, the existing subject matter theory is not well developed. Criminology is no exception. Many of the key variables have been identified, but how they are related to one another is either unknown or known only up to the sign of the relationship. When functions are specified, they are assumed to be relatively simple. Linear functions and log functions are common. If

---

[8] Following Hastie et al. (*2009:298*), suppose there are $t = 1, \ldots, T$ trees and $i = 1, \ldots, N$ observations (e.g., 50 trees and 500 observations). We are seeking the set of $\hat{f}_t$ functions linking each of, say, 50 predictor vectors to the response.

1. Initialize: $\hat{\alpha} = \mathrm{prop}(y_i), f_t = f^0, t = 1, \ldots, T$. The value of α is the response variable proportion over all observations. This does not change. The functions for each predicator are initially set to zero.
2. Cycle: $t = 1, \ldots, T, 1, \ldots, T, \ldots$

$$\hat{f}_t \leftarrow \mathbf{S}_t(y_i - \hat{\alpha} - \textstyle\sum_{r \neq t} \hat{f}_t(x_{it})),$$

$$\hat{f}_t \leftarrow \hat{f}_t - \tfrac{1}{N} \textstyle\sum_{i=1}^{N} \hat{f}_t(x_{it}),$$

where $\mathbf{S}_t$ is a smoother. Continue #2 until the individual functions do not change. The cycling depends on constructing a long sequence of pseudo residuals and fitting those with a smoother such as smoothing splines or lowess.

[9] Work is underway on theory that might be used for multinomial outcomes. In principle, the prior tree distribution could be altered to allow for asymmetric costs.

the research goal is to broadly characterize the most important relationships in an easily-understood manner, rough approximations can be useful.

But when forecasting accuracy is the goal, simplicity can have a very high price. Important associations will sometimes be highly nonlinear. In addition, even when the most important variables have been identified, there may be a large number of other variables, often with marginal substantive roles, that nevertheless each carry useful predictive information for small subsets of cases. In other words, there can be a large number of small associations of little explanatory interest, that in the aggregate can improve forecasting accuracy. Finally, these small relationships do not have to be represented by interpretable measures. Surrogates, often distant surrogates, will suffice for forecasting even though they cannot be anticipated by subject-matter experts and would likely be of little interest in any case.

Ensembles of classification trees are well suited for this setting for the following reasons:

1. There is no need for a model of the data generation process beyond random realizations from a joint distribution of the response and the predictors. Many of the most common concerns in social science models are irrelevant. For example, uncertainty is produced by the theoretical equivalent of probability sampling. There is no disturbance term whose properties are so critical in conventional regression models.
2. There is no need for cause and effect, and there is no reliance on causal models. This is good because machine learning procedures are not causal models. Their primary purpose here is forecasting for which cause and effect can be irrelevant.
3. By using discretized variables with each tree, unanticipated and highly nonlinear functions can be empirically approximated in a manner that can be very effective in reducing bias in fitted values. There is no need to specify functional forms in advance.
4. The stagewise approach used in classification trees, in which earlier partitions are maintained, can easily lead to unanticipated interaction effects inductively derived that can reduce bias. Many of these interactions may be of a very high order that researchers rarely consider. Often these will be effective surrogates for predictors not directly measured. For example, a high-order interaction effect between gender, age, and zipcode may be a good proxy for gang membership. Gang membership does not have to be explicitly included among the set of predictors.
5. Generating a large number of different trees from many passes through the data helps distinguish features of the data that are systematic from features of the data that are happenstance. For example, relationships found very rarely are likely to be treated as noise.
6. The training data can be explored from many different vantage points using many different variants of the exploration tools. A relationship that may seem unimportant in the training data as a whole, may turn out to be very important when situated in a different sample and characterized by a different collection of predictors.

7. Aggregating over trees can smooth the step functions produced by each tree when that is appropriate. In criminal justice applications at least, quantitative predictors are rarely associated with the response as step functions.
8. Aggregating the results from a large number of classification trees will introduce stability not obtained from a single tree. Forecasting variance will be reduced.
9. There will often be very instructive output beyond measures of procedure and forecasting performance: measures of forecasting importance for each predictor and graphs of how each predictor is related to the response.

In summary, when subject-matter theory is well-developed and the key predictors are included in the training data, conventional parametric regression methods can forecast well. There is probably no need for the sorts of procedures emphasized here. However, when the subject-matter theory is under-developed and important predictors are not available, tree ensembles (and other forms of machine learning) can can offer forecasting skill well beyond that of conventional social science approaches.[10]

Perhaps the key asset of tree ensembles is that they take very seriously the goal of searching for structure. The sample space defined by $Pr(G, X^*)$ can be explored from many different vantage points, using many different variants of the fitting procedures. In random forests, the random samples of data provide different views of the sample space, and the samples of predictors provide different fitting opportunities. Stochastic gradient boosting gets much the same job done by sampling the training data and reweighing the sample with each pass. Bayesian trees relies on samples of parameter values so that a heterogenous mix of trees is applied to the data. In the end, all three approaches combine the output across trees so that a useful balance between the bias and the variance may be achieved.

---

[10] Another very good machine learning candidate is support vector machines. There is no ensemble of trees. Other means are employed to explore the sample space effectively. Hastie et al., (*2009: 417-436*) provide an excellent overview. What experience there is comparing support vectors machines to the tree-based methods emphasized here is that usually they all perform about the same. When they do not, the differences are typically too small to matter much in practice or are related to unusual features of the data or simulation.

Chapter 6
Examples

Abstract In order to help illustrate the ideas from previous chapters, this chapter provides detailed examples of criminal justice forecasting. These are real applications that led to procedures adopted by criminal justice agencies. As such, they combine a number of technical matters with the practical realities of criminal justice decisions-making. For the reasons already addressed, random forests will be the machine learning method of choice.

6.1 A Simplified Example

We now turn to some examples of real forecasting with real data. The goal is to help fix the key ideas from the previous chapters. We begin with an example that is simplified for didactic purposes. In particular, only five predictors from a much larger set are used. A much more complicated forecasting enterprise will be considered later.

The forecasting task in this example was to predict which individuals on parole or probation (largely probation) in an East Coast city "failed" by committing a homicide or attempted homicide or by being the victim of a homicide during an 18 month period beginning at intake. Any other outcomes, including commission of other crimes, were not failures. The forecasts were to be used in a homicide-prevention program designed to reduce homicides in which the individuals under supervision were involved, either as perpetrators or victims.

It might seems strange from criminological point of view to combine in one outcome homicide perpetration and homicide victimization. But local craft lore and an initial analysis of the data indicated that homicide perpetrators and homicide victims were, in this setting, largely the same kinds of individuals. Commonly, they were men under 25 with arrests for very serious crimes beginning in their early teens, who also had easy access to handguns. One "disrespect" too many could rapidly lead to violence in which a single individual might as easily be the shooter as the person shot.[1]

Predictors were drawn from the usual sorts of administrative records routinely available in electronic form and projected to be available in real time when subsequently actual forecasts would need to be made. Predictors included prior record, age, the instant crime of conviction, gender, age, the age of an offender's first arrest, and many others. There were certainly no surprises.

Several different cost ratios of false negatives to false positives were applied, all capturing the far greater relative cost of false negatives. False negatives were defined as perpetrators of a homicide or attempted homicide or victims of homicide who were not forecasted as such. False positives were defined as individuals who were incorrectly forecasted to fail as perpetrators or victims.

Table 6.1 below shows the forecasting results when the cost ratio was set at 15 to 1 for the cost of false negatives to the cost of false positives.[2] For the random forests program in R (*randomForest*), this was done using its stratified sampling procedure. There were two strata, one for those who failed and one for those who did not. Each time a random sample of observations was chosen, 200 of the former and 500 of the latter were chosen (i.e. *sampsize* $= c(500, 200)$.[3] This led to a ratio in the table of 1535 false positives to 99 false negatives, or about 15.5.

There is no way to directly impose a given cost ratio on the confusion table. Good practice seems to begin by using a sample of about two-thirds for the category with the smaller number of cases, and then adjusting the sample size for the other category until the desired cost ratio in the table is approximated. Starting with the two-thirds sample for the less common outcome, leaves a sufficient number of out-of-bag (OOB) cases for test data, and that is the main point. One could work with 50% or 75% sample, if that is required to arrive at the target cost ratio. Settling on a satisfactory cost ratio in a confusion table involves some trial and error, but the process is useful to get a sense of how different cost ratios affect performance. In this instance, the correspondence between the target of 15 to 1 and the empirical results is high. From the two off-diagonal cells, $1536/99 = 15.5$. The forecasting results were similar with cost ratios as small as 10 to 1 and as large as 20 to 1.

---

[1] This pattern is common. In Philadelphia, for example, "Last year, 85 percent of the city's homicides were African American, almost all of them male. Four of five killers were African American males, demographically indistinguishable from their victims." ... Quoting Mayor Michael Nutter: "The No. 1 issue for homicide in Philadelphia is generally classified as an 'argument.' " (*Heller, 2012*).

[2] For this analysis, the code in the random forests program in R (*randomForest*) was written as: $rf3 \leftarrow randomForest(morefail \sim iassault + igun + priors + intage + sex, data = temp1, importance = T, sampsize = c(500, 200))$. The output was saved under the name of *rf3*. The variable *morefail* was the response. There were 6 predictors starting with *iassault* and ending with *sex*. The input data were *temp1*, predictor importance plots are requested with *importance=T* and *sampsize* = *c(500,200)* determined the sampling strategy for each tree. The assignment symbol $\leftarrow$ is produced in R by a $<$ followed immediately by a $-$.

[3] The order in $c(500, 200)$ is numerically (low to high) or alphabetically depending on how the two outcome categories are coded.

	Forecasted Not Fail	Forecast Fail	Model Accuracy
Not Fail	12674	1536	.89
Fail	99	153	.61
Forecasting Accuracy	.99	.10	

Table 6.1 Confusion table for forecasts of homicide or attempted homicide perpetrators and homicide victims for a 15 to 1 cost ratio. True negatives are identified with 89% accuracy. True positives are identified with 61% accuracy. Non-failures are forecasted with 99% accuracy, and failures are forecasted with 10% accuracy. (Fail = perpetrator or victim. Not Fail = not a perpetrator or victim.)

The main purpose of confusion tables is to determine forecasting skill. Because a proper confusion table is constructed with OOB data, true forecasts are made. But forecasting skill can be examined in several different ways.

One might think that the overall proportion of cases correctly classified (i.e., the sum of the counts in the two main-diagonal cells divided by the total number of observations) is the primary measure of forecasting accuracy. But it is not. Recall the earlier discussion of uncertainty. The overall proportion correctly classified weights false negatives and false positives equally. The weights should represent relative costs and here at least, they are explicitly not equal. Moreover, the overall proportion is insensitive to how the forecasts might actually be used. The overall proportion conditions on neither the actual outcome or the forecasted outcome.

A better option is to consider how well the procedure performs when the truth is known. Given a known outcome (e.g., homicide perpetrator or victim), what proportion of the time does the procedure identify cases correctly? Substantial accuracy can indicate that the forecasting procedure is able find what it was designed to find. This is an encouraging start. The procedure has promise. But, it does not directly address forecasting accuracy.

For forecasting accuracy, one should condition on the *forecast*. Given a forecast (i.e., a forecast of homicide perpetration or victimization), what proportion of the time is it correct? By conditioning on the forecast, there are close parallels to how a forecasting procedure will be applied in practice when criminal justice decision makers try to anticipate possible threats to public safety. Substantial accuracy indicates that in actual use, the forecasting procedure could be helpful.

However, forecasting performance measures are necessarily matters of degree. It can be very important, therefore, to have baselines from which to make comparisons. A good baseline is the proportion of times a forecast would be correct if none of the information in the predictors were used. Let's play this through.

In this example, about 2% of the individuals fail. So, if non-failure is forecasted, it would be correct approximately 98% of the time. If failure is forecasted, it would be correct approximately 2% of the time. So why not ignore the predictors and always forecast non-failure? It is difficult to imagine a forecasting procedure that in this setting could have better than 98% accuracy. The answer is that the 2% of cases that become false negatives are very costly. Given the 15 to 1 cost ratio, decision-makers are prepared to accept a substantial increase in the number of false positives if the number of false negatives can be meaningfully reduced. They prefer to have *less* accurate forecasting overall if it means avoiding especially costly forecasting errors. This underscores why costs must be built into forecasts if those forecasts are to be responsive to real policy considerations.[4]

How does all this work out here? Consider first the rows in Table 6.1. The row proportions show how good the forecasts are when the actual outcomes in the test data are known.[5] Of those known not to fail, random forests gets it right about 89% of at the time. Of those known to fail, random forests gets it right about 61% of the time; a substantial majority of rare events are being correctly identified.[6]

Consider now the columns from which one can compute how often the forecasts are correct. That accuracy depends in part on the false-negative to false-positive cost ratio. In this instance, decision-makers are prepared to accept a large number of false positives, which implies that a substantial number of forecasting errors will be made when failure is the forecast. This is not a weakness in the forecasting approach. It is a strength. The risk preferences of decision-makers are explicit and taken seriously.

When no failure is forecasted, that forecast is correct 99% of the time. This is a slight improvement over the baseline of about 98%. When failure is forecasted, that forecast is correct 10% of the time. On its face, this may be disappointing, but it results substantially from the large number of false positives that decision-makers are prepared to tolerate. There are about 10 false positives for every true positive. Moreover, the 10% figure is five times larger than the base rate of 2%. In short, using random forecasts with only five predictors is in this example at least encouraging.

Another important part of the random forests output is a plot of predictor importance when the response outcome is known. The focus is on the rows of a confusion table. From Figure 6.1 one can see that gender is the most important predictor for finding high risk cases. If its values are shuffled, forecasting accuracy for a failure drops by over 30 percentage points. The 61% accuracy is now less than 30%. Similar calculations can be made for the other predictors and because the outcome is binary, the importance plot for the "successes" would convey the same information.[7]

The forecasting importance of gender can be used to illustrate the tension between finding subgroups with high concentrations of high risk individuals and legal constraints. On the one hand, gender can be a "suspect" class in civil litigation over discrimination. More broadly, one could wonder about the justice of treating "similarly situated" men and women differently. On the other hand, dropping gender from the set of predictors could mean that half of the potential perpetrators and victims who would be correctly identified in advance, might not be unless one could find an *acceptable* surrogate predictor for gender. Given the large number of pro-

---

[4] In effect, the false negative 2% might be considered a false negative 30% (2% × 15) when relative costs are factored in.

[5] For ease of discussion, the proportions on the right and bottom margins represent the proportion correct, not proportion incorrect. We proceed in this manner in all subsequent confusion tables as well, despite the more common practice of reporting the proportion incorrect.

[6] Using a larger number of predictors, random forests correctly identifies failures about 80% of the time.

[7] An "immediate" crime is the crime of the most recent conviction after which the probation or the parole decision was made.

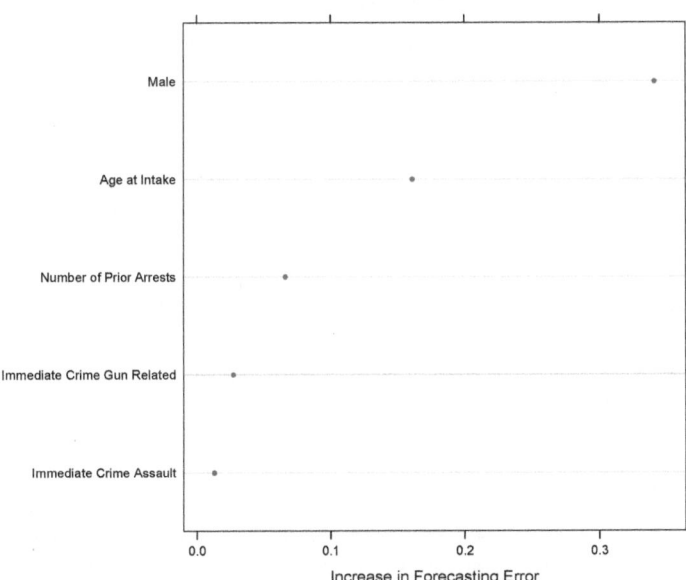

Fig. 6.1 Predictor importance measured by proportional reductions in forecasting accuracy for perpetration of or victimization from homicide while under supervision.

bation/parole cases in this city, the decline in forecasting accuracy translates into roughly 100 homicides or attempted homicides. A decision about whether to include gender as a predictor can have real consequences.[8]

There is also for each predictor a plot of the response functions. We consider two such plots as illustrative. Figure 6.2 shows the partial response plot for age. One can see that the risk of failure declines sharply with age until around age 40. There is then a modest increase until about age 60. After that, there are too few observations to reveal a pattern. The decline from the teenage years to early middle age is consistent with decades of research, although too often the decline is represented very crudely (e.g., under 30 versus over 30). Then, information that can be important for forecasting is lost. The increase after age 50 is a surprise and may represent domestic violence incidents.

Each value on the vertical axis is a centered logit value of $\log[p/(1-p)]$. One can solve for p for different logit values on the vertical axis. In a probability metric, the affect of age is dramatic. The probability of failure drops by about .50 when an 18-year-old is compared to a 40-year-old.

What about gender? Figure 6.3 shows the partial response plot. Not surprisingly, men are more likely to fail. Their probability is larger by about .06. Despite the

---

[8] There may well be no such thing even in principle as an acceptable surrogate, unless the primary concern is public relations. Consequently, one is again faced with difficult tradeoffs. What price in units of potential victimizations can one tolerate for using predictors that are less troubling?

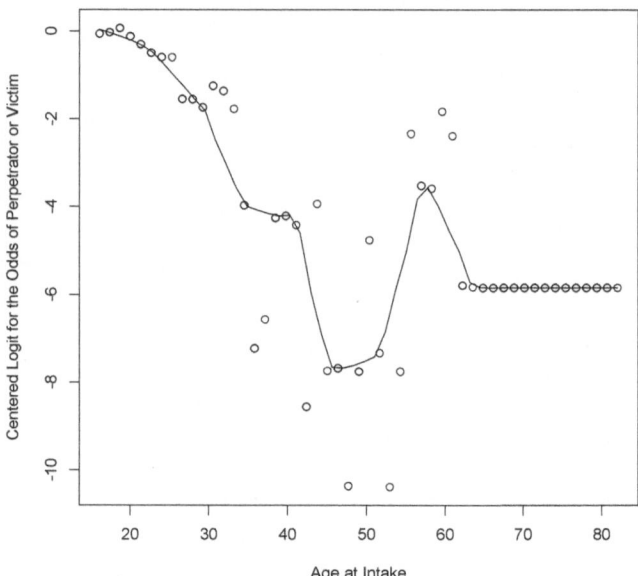

Fig. 6.2 Partial response plot for "failure" against for age. The vertical axis is in centered logits. The relationship is highly nonlinear.

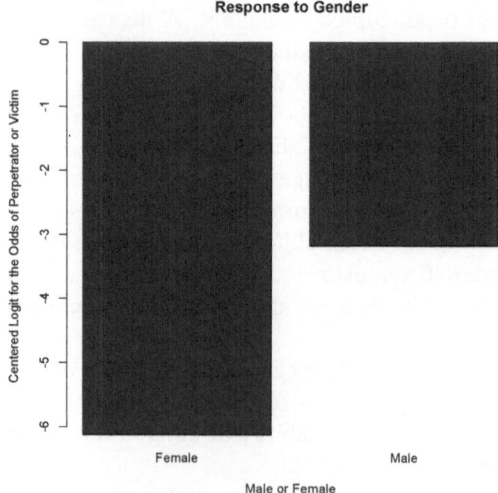

Fig. 6.3 Partial response plot "failure" against gender. The vertical axis is centered logits. Men are more likely to fail.

dominant role of gender for predictor accuracy, the difference between men and women in the probability of failure might seem relatively modest. But there is really no tension between the two. The partial response does not directly take into account the variability in the predictor unrelated to other predictors. Forecasting accuracy does. So a modest partial response can be coupled with relatively large forecasting importance if the predictor has a relatively large residual variance after adjustments are made for its relationships with other predictors. [9]

In short, a partial response plot and forecasting importance address different questions. Only a bit too simply, a partial response shows on the average how much the outcome changes depending on the values of the predictor. Forecasting importance translates the partial response into forecasting accuracy that depends on the heterogeneity in those values.

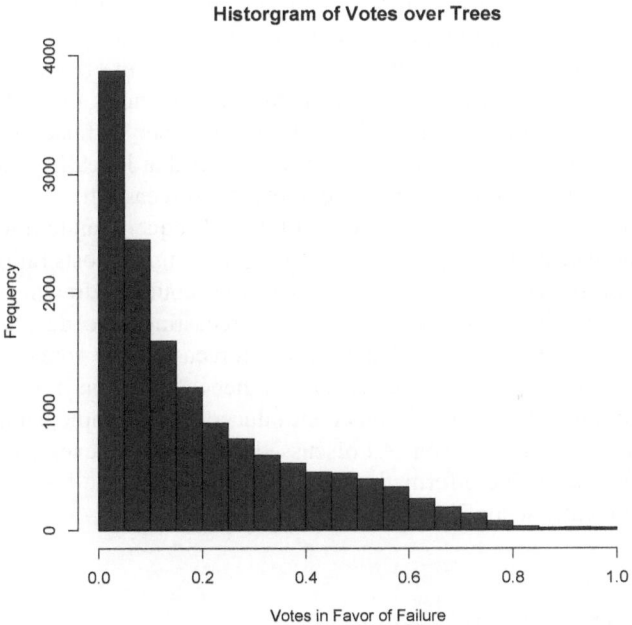

Fig. 6.4 A histogram of the proportions (over trees) of votes for failure. There is one proportion for each case. Cases with proportions near .50 are the cases for which the forecasts have the greatest uncertainty.

Finally, there is the matter of forecasting uncertainty. One useful approach is to examine the distribution of votes over trees for each forecasted class. For insight

---

[9] This reasoning is much like the reasoning that applies to the difference between a partial regression coefficient and the "explained" variance uniquely attributable to a predictor.

about forecasting uncertainty in the procedure overall, a histogram of the votes can be instructive.

Figure 6.4 is a histogram of the votes for a failure over all of the trees in the random forecast based on the OOB data for 14,462 observations. The horizontal axis shows the proportion of votes in favor of predicting a failure. The vertical axis shows the number of cases. The skew to the right is dramatic. Most of the forecasts are for no failure, often with strong majorities.

All of the bins to the right of .50 include cases for which a failure is forecast. About half of the vote proportions greater than .5 are less than .60 (i.e., 786). There is usually no obvious threshold for an unacceptable amount of uncertainty, but when the vote proportions are near .50, the forecasts are not much more reliable than a coin flip. Here, that might be cases with votes in the .50 to .60 range.

One might consider simply throwing out all cases for which the vote is too close to .5, and then recompute the confusion table. The table would show forecasting accuracy for those cases that were forecasted in a sufficiently reliable manner. Some might feel that this is a more instructive way to represent model performance.

However, by raising the threshold (e.g., to .60), one is, in effect, shrinking the relative costs of false negatives to false positives. The intent is to be "more sure" before a forecast of failure is made. In trade, the number of false positives will decrease as the number of false negatives increases. If that is really how decision-makers want to proceed, they should recompute the forecasts from the beginning using a new cost ratio. If that is not done, all of the subsequent random forests (e.g., predictor importance plots) output will still reflect the original costs ratio when that is no longer appropriate. Very misleading conclusions could be drawn.

A second approach can be useful when the forecasting procedure has become operational. As the outcome for each new case is forecasted, the vote over trees can be reported. When the vote is too close to .5, other information *not already used* when the random forest was grown can be introduced. For example, an inmate may have a job waiting outside prison. As discussed earlier, it is generally a bad idea to override a forecast using information already contained in the forecasts. Human judgement is typically second best.

6.2 A More Complex Example

Consider now a more elaborate application. For an agency that supervised individuals on probation and parole, the policy question was how best to allocate supervisory resources. Given tight budget constraints, the goal was to make the intensity and content of supervision cost-effective. Without sacrificing public safety, more expensive supervision would be provided only to offenders most in need.

Three classes of offenders were proposed: (1) individuals committing a serious crime, (2) individuals committing a crime but not a serious crime, and (3) individuals not committing any crime. "Serious" crimes were essentially Part I crimes, and crimes that were not serious were essentially other than Part I crimes.[10] The distinction between crimes that were serious and those that were not was partly a matter of practicality and political sensitivities. Serious crimes could not be defined so broadly that agency resources would be rapidly exhausted. At the same time, certain crimes that stoked public anxieties and that got the attention of local public officials had to be included.

There were also statistical considerations in how the outcomes were defined. In general, the more heterogenous an outcome class, the more difficult it can be to arrive at accurate forecasts. When the goal is to forecast different kinds of criminal behavior, one needs to think hard about crime etiology — insofar as credible information exist — because both the predictors and how they function can differ by crime type. To the degree that offenders "age-out" of different kinds of criminal activity at different ages, for instance, having many different kinds of crime in a given outcome category will tend to dilute the impact of age and reduce the accuracy of any forecasts that depend on age-related effects.

It follows that there can be tensions between forecasting accuracy and the "real politik" of supervisory decisions. For example, should rape have been put in the serious-crime outcome category along with armed robbery? In this instance, political considerations trumped statistical considerations, and it was. For several other less salient kinds of crime, statistical considerations won the day.

The vast majority of offender types were included in the forecasting exercise. But, individuals identified from their prior record or current conviction as "sex offenders" were, by and large, assigned automatically to special supervisory services. No forecasting was undertaken. There were analogous procedures for offenders with significant histories of drug use or major psychological problems. Finally, a relatively small number offenders were diverted to experimental, community-based, programs without the use of any forecasts. In short, allowance was made for a variety of placement practices, many of which were not informed by forecasts. Indeed, policy considerations sometimes precluded using forecasts to inform placement decisions. Whether forecasts might have been helpful for other purposes was not considered. For example, higher risk drug offenders might have been subject to more frequent testing.

Commission of a crime was operationalized as an arrest within two years of intake. Training data included over 119,000 cases from a three-year intake window. That is, all new cases during those three years were included. For those data, 9.6% failed with an arrest for a serious crime, 30.9% failed with an arrest for a crime that was not defined as serious, and 59.3% did not fail. Predictors were taken from administrative records routinely available in electronic form. Random forests was applied.[11] Table 6.2 shows the resulting confusion table.

---

[10] Serious crimes included murder, attempted murder, aggravated assault, robbery or a sexual crime.

[11] $output \leftarrow randomForest(threeway \sim iseriouscount + Asexpriors + Jfirstage + seriousyears + Aviolpriors + jailpriors + Afirstviolage + age + Afirstage + jaildaypriors + Aallpriors + Zipfac, data = w2, importance = T, sampsize = c(10000, 10000, 10000))$.

	Forecast Serious Crime	Forecast No Crime	Forecast Not Serious	Model Accuracy
Serious Crime	7112	2220	2248	.62
No Crime	7147	49187	14867	.69
Not Serious	4553	9601	23000	.62
Forecasting Accuracy	.37	.80	.57	

Table 6.2 Confusion table for forecasts of parole or probation performance: arrest for a serious crime, arrest for a crime that is not serious, and no arrests. Arrests for serious crimes are identified with 62% accuracy, arrests for crimes that are not serious are identified with 69% accuracy, no arrests are identified with 62% accuracy.

Before digging into the full confusion table, we need to revisit the matter asymmetric costs. With three outcomes introducing asymmetric costs can get tricky. Tables 6.3 though 6.5 are 2×2 subtables from Table 6.2, constructed for ease of exposition.

	Forecasted Serious Crime	Forecast No Crime
Serious Crime	7112	2220
No Crime	7147	49187

Table 6.3 Confusion table for forecasts of parole or probation performance: arrest for a serious crime and no arrests. The cost ratio is about 3.2 to 1 for false negatives to false positives, where serious crime is a positive and no crime is a negative.

	Forecasted Serious Crime	Forecast Not serious
Serious Crime	7112	2248
Not Serious	4553	23000

Table 6.4 Confusion table for forecasts of parole or probation performance: arrest for a serious crime and an arrest for a crime that is not serious. The cost ratio is about 2.0 to 1 for false negatives to false positives, where serious crime is a positive and crime that is not serious is a negative.

	Forecasted No Crime	Forecast Not Serious
No Crime	49187	14867
Not Serious	9601	23000

Table 6.5 Confusion table for forecasts of parole or probation performance: arrest for a crime that is not serious and no arrests. The cost ratio is about 1.5 to 1 for false negatives to false positives, where a crime that is not serious is a positive and no crime is a negative.

Table 6.3 is the confusion table for arrests for serious crimes and no arrests at all. Serious crime is taken as a positive, and no crime is taken as a negative. The intent was to make false negatives about 10 times more costly than false positives. In reality, the ratio is about 3.2 to 1 (7147/2220).

Table 6.4 shows the confusion table for arrests for serious crime and for crimes that were not serious. Serious crime is taken as a positive, and crime that is not serious is taken as a negative. The intent was to make false negatives 5 times more costly than false positives. In fact, the ratio is about 2 to 1 (4553/2248).

Table 6.5 shows the confusion table for no arrests and arrest for crimes that were not serious. Non-serious crime is taken as a positive and the absence if crime are taken as a negative. The intent was to make false negatives 5 times more costly than false positives. In fact, the ratio is about 1.5 to 1 (14867/9601).

By fine-tuning the stratified sampling turning parameters, it would have been possible to arrive cost ratios quite close to their targets. But decision-makers imposed two constraints on the forecasts that had important implications for costs and forecasting accuracy. First, the fraction of cases forecasted to be arrest-free but then arrested for a serious crime — the worst kind of forecasting error — had to be well under 10%. There was nothing special about the 10% figure. It was an approximation of what decision-makers thought would be acceptable to key stakeholders and the public.

Second, the fraction of individuals forecasted to have a serious arrest could not exceed about 17%. Individuals identified as a serious threat to public safety were to be assigned to intensive supervision. Intensive supervision is very expensive, and as a fiscal matter, there was an upper bound to the number of very high risk cases that could be intensively supervised. Like so many public agencies, these decision-makers were being asked to do more with less.

Further complicating matters was that one could reduce the first percentage only by increasing the second. For example, in order to reduce the number of worst-case forecasting errors, one could accept a lower standard of evidence for an individual forecasted to be arrested for a serious crime. But that would increase the number of false positives and the overall number of those forecasted to commit a serious crime. The 17% upper bound might then be exceeded.

Responding sensibly to these constraints meant significant compromises in the initial cost ratios. As a matter of arithmetic, the desired costs ratios were incompatible with the imposed constraints. Something had to give. In this case, decision makers had to settle for smaller cost ratios than they initially preferred.

Ideally, one would want to rethink the cost ratios trying to incorporate each constraint's cost implications. For example, the costs of false positives for serious crimes increase substantially as the upper bound of 17% is approached. Unfortunately, we could not arrive at a practical way to formally build in cost ratios that changed depending on the forecasting outcomes. It was the classic chicken and egg problem.

Nevertheless, the relative costs of different kinds of forecasting errors were necessarily a key driver that could be manipulated. One could reduce the number of individuals forecasted to be crime free, but who actually were arrested for a serious crime, by increasing the relative costs of such false negatives. One could reduce the number of individuals forecasted to commit serious crimes by decreasing the relative costs of both kinds of false negatives (i.e., with respect to forecasts of no crime and forecasts of crimes that were not serious). In short, there were things that could be done, and the goal was to stay within the two constraints while producing sufficiently accurate forecasts.

As before, we manipulated the number of cases sampled within each of the outcome classes. About two-thirds of the violent crime cases were sampled. The number of cases sampled for the other two outcome classes was determined by trial and error. Eventually, sampling 10,000 of each outcome class led to an acceptable result.[12]

In the end, 3.6% of the individuals forecasted to not be arrested for any crime were arrested for violent crimes. That percentage was well below the target of 10%. The percentage could have been substantially smaller were it not for the second requirement (i.e. that no more than 17% could be forecasted to be commit a serious crime). We settled on a total of 15.5% projected to be arrested for a serious crime. In short, the targets for both constraints were met. One might even argue that the cost ratios in Table 6.2, which are very different from their initial targets, better reflect the relative costs of different forecast errors because they incorporate the two outcome constraints.

Consider now some key details of Table 6.2. With three outcome classes, there are 9 cells with counts. The numbers of correct forecasts are shown along the main diagonal. As before, model accuracy is reported on the right margin of the table, and foresting accuracy is shown on the bottom margin of the table.

Model accuracy was acceptable to decision-makers. Around two-thirds of individuals for each of the three outcomes are correctly identified. One of the two-thirds figures was for some especially impressive. For the full set of cases, about 6% are arrested for serious crime. Serious crimes were relatively rare. Yet, within the subset of individuals who actually are arrested for a serious crime, 62% are correctly identified by random forests.

Forecasting accuracy looks good as well. When no crime is forecasted, the forecast is correct 80% of the time. As just noted, the worse-case forecasting error occurs only 3.6% of the time. A strong argument might be made for employing a form of less intensive supervisory if an individual is forecasted to be crime-free.

When an arrest for a crime that is not serious is forecasted, the forecast is correct 57% of the time. The vast majority of the forecasting errors are for individuals who were crime-free. Only 13% of those incorrectly forecasted were arrested for a serious crime.

When a serious crime is forecasted, the forecast is correct 37% of the time. About 40% of the forecasting errors are for those arrested for a crime that was not serious and about 60% are for those who were crime-free. The 40% figure has some positive implications because these individuals were engaged in criminal behavior even if the arrest was not for a serious crime. The 60% figure is more troubling. But it is also

---

[12] Recall that the sample sizes alter the prior distribution of the response for each tree, which in turn alters to loss associated with each kind of forecasting error.

likely that at least some of these individuals were false negatives in another sense: they were not being caught for the crimes they were committing. In any case, all of these numbers could have been rather different, and arguably better, were it not for the two constraints. But as required, the total does not exceed the 17% upper bound.

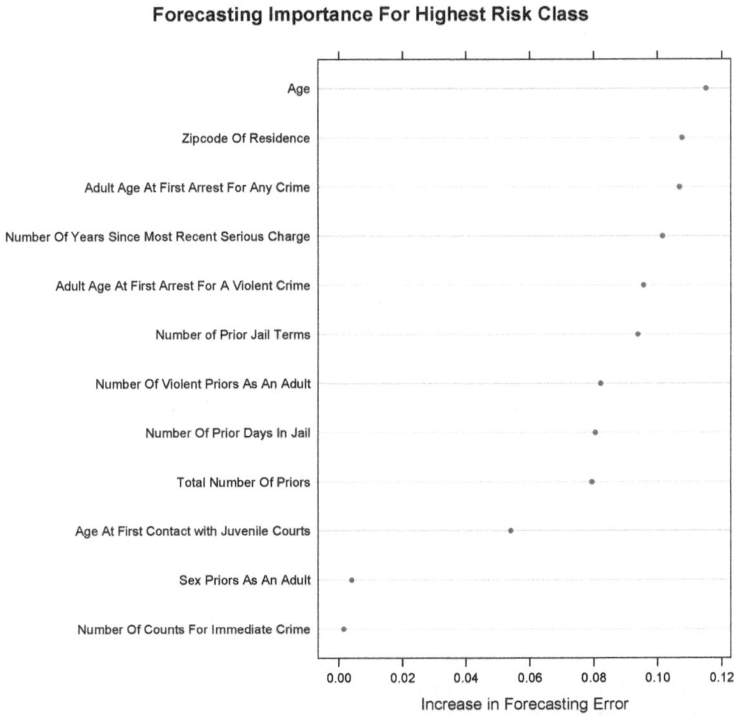

Forecasting Importance For Highest Risk Class

Fig. 6.5 Reductions in forecasting accuracy for each predictor after shuffling with respect to arrests for a violent crime. Predictors are ranked by forecasting importance.

Measures of forecasting importance are also more complicated when there are more than two response categories. The key question is: forecasting importance of what? There are two common options. One can evaluate the forecasting importance of predictors for a given outcome class, or one can evaluate the average forecasting importance across all outcome classes. Both can be useful.

Figure 6.5 shows the forecasting importance plot for a single outcome class: an arrest for a serious crime. One can see, for example, that if the age variable is shuffled, forecasting accuracy for serious crimes declines by about 12 percentage points. Almost as important are the zip code of residence, the earliest adult age for an arrest, and the number of years since the most recent serious charge. It does not matter much if a person has a sex offense prior or has a large number of counts for their current crime of conviction.[13]

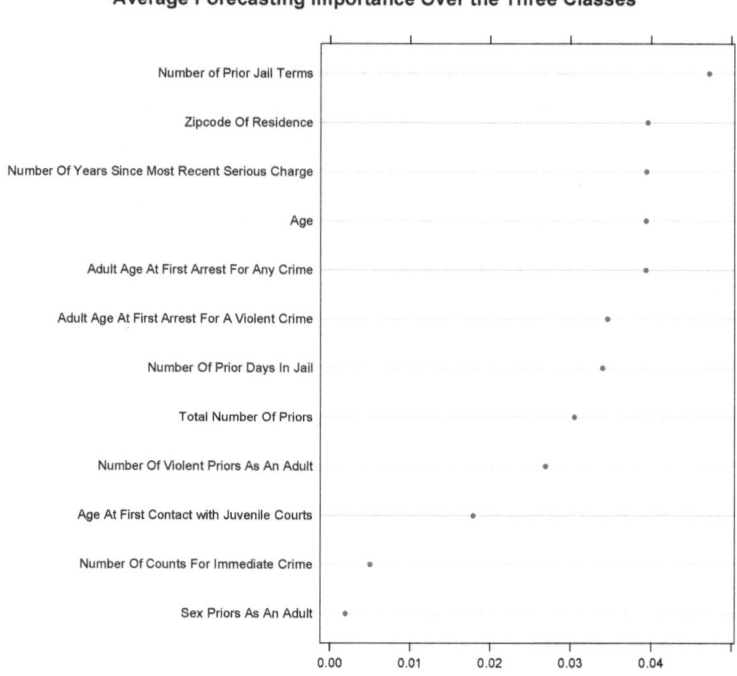

Fig. 6.6 Average reductions in forecasting accuracy for each predictor after shuffling with respect to all three outcomes. Predictors are ranked by forecasting importance.

Figure 6.6 shows average forecasting importance for all three outcome classes at once. That is, the forecast importance for each class is summed and then divided by number of classes. The number of prior jail terms now rises to the top followed by the four most important predictors from Figure 6.5.

Average forecasting importance is smaller than forecasting importance for serious crimes because for some outcome categories, it is more difficult to have large forecasting contributions. In particular, it is more difficult to have substantial forecasting importance for outcome categories that contain a larger number of cases. For example, suppose that for two different outcome categories, each has a forecasting accuracy of 50%. And in both cases, suppose shuffling a given variable increases the number of forecasting errors by 25. If there are 100 cases in one of the outcome categories, 50% accuracy drops to 25%. If there are 500 cases in the other outcome

---

[13] Having a prior for a sex offense does not necessarily mean that the individual is labeled a sex offender. The sexual offense may be minor, in the distant past, and unrelated to current criminal behavior.

category, accuracy drops to 45%. In this application, there are many more observations in the non-serious crime and no-crime categories than in the serious crime category. So, forecasting importance tends to be substantially smaller predictor-by-predictor for the two happier outcomes. This leads to smaller predictor importance averaged over the three outcomes. One implication is that the forecasting importance for violence crime tends to dominate average importance, which helps explain why the most important predictors in the Table 6.5 and Table 6.6 are similar.

Forecasting importance has no direct implications for forecasting accuracy. It comes into play primarily as decision-makers try to understand what the forecasts mean and/or convince others that the forecasts have real benefits. The most important predictors should "make sense." They typically do if drawn intelligently from existing administrative records. The surprises can usually be "explained" with a little *post hoc* reasoning. More problematic are predictors thought to be important for forecasting, but turn out not to be. Such findings can challenge long-accepted craft lore, past research in peer reviewed journals, and well-respected professional reputations. Perhaps the best response is to emphasize that the task at hand is forecasting, not explanation, and that because many popular predictors can be highly correlated, some may mask the performance of others. There is also forecasting skill that cannot be attributed to any single predictor, but that may be responsible for "null effects."[14]

Forecasting importance can also play a key role when decisions are made to exclude certain predictors on legal or ethical grounds. For example, if race has little forecasting importance when included with other predictors, it can be dropped from the training data should there be even a hint of concern. However, if race has substantial forecasting importance when included with other predictors, significant tradeoffs can arise. How much forecasting accuracy are decision-makers prepared to forego if race is no longer included? That, in turn, requires that decision-makers think through the consequences of different forecasting errors.

Suppose, for instance, that excluding race as a predictor will plausibly lead to 5 more homicides. Because, as noted above, people tend to kill people like themselves, there could be 5 more homicides of individuals with the same general background as the individuals whose behavior is being forecasted. Perhaps 40 young males from certain neighborhoods will not have to live with intensive oversight, but the price could be 5 young male murder victims from those same neighborhoods.

In addition to knowing the forecasting importance of each predictor, it can be useful to learn how each predictor is related to the response. There is no such thing as an average response function over response categories. Response functions are constructed separately for each predictor and each response category. But just like with forecasting importance, the response function should "make sense."

Recall how the response is represented. It is the difference between the log of the odds for the outcome in question and the log of the odds averaged over all outcome

---

[14] That said, it often seems that when machine learning is applied in criminal justice forecasting, the stronger predictors tend to be behavioral, not attitudinal. Various features of anti-social behavior in the past can be excellent predictors of anti-social behavior in the future. Psychological profiles or inferences about states of mind typically add little to forecasting accuracy once the behavioral measures are included.

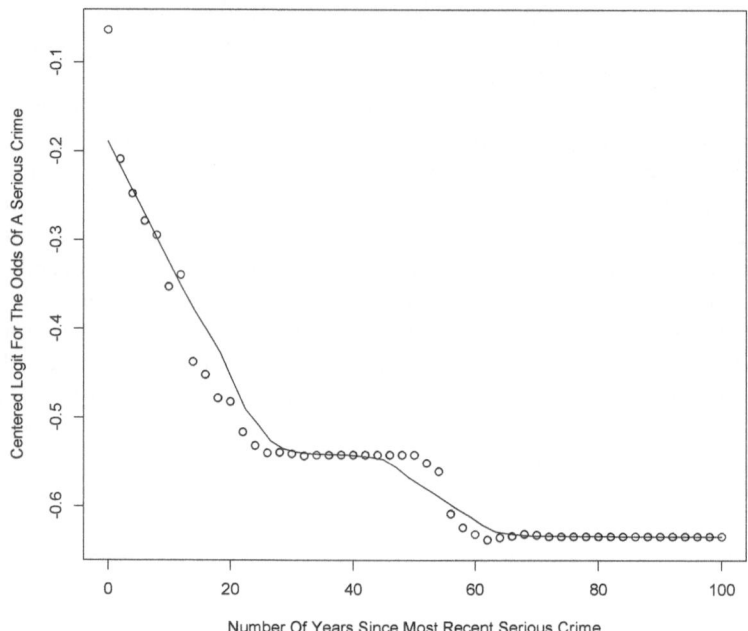

Fig. 6.7 Response of an arrest for a violent crime to the number of years since the most recent serious crime. A value of 100 is used for individuals with no serious prior arrests. There is little data beyond a value of about 50.

classes. In other words, it is measure of disparity between the response for a given outcome class and the average response over all outcome classes. With our three crime-related outcomes, it is the difference between the response in log-odds units for, say, violent crime and the average response in log-odds units over all three crime-related categories.

When there are only two response classes, partial response plots are relatively easy to interpret. The response plot for one response class is just the mirror image of the response plot for the other response class. One of the two is redundant.

When there are more than two response classes, as in this example, response plots are more challenging. There is information in the partial response plot for each response class. If there are, say, three classes, there is a need for three response plots. There is far more output to digest because no single plot is redundant.

Figure 6.7 shows the partial response plot for the serious crime outcome and the predictor years since the most recent serious crime. One should anticipate a negative relationship but after that, the criminology literature provides little guidance. In fact, the relationship is steeply negative until a value of about 20 years is reached. It levels off. Broadly, the response function makes sense. Consider how credible the overall results would be if the relationship was strongly positive.[15]

As noted earlier, the centered logits can be transformed into probabilities. When that is done here, the relationship is non-trivial. If one compares the probability of being arrested for a serious crime for someone whose past serious crimes were a year old or less to the probability of being arrested for a serious crime for someone whose past serious crimes were 4 decades earlier, probability of failure declines by about .13.

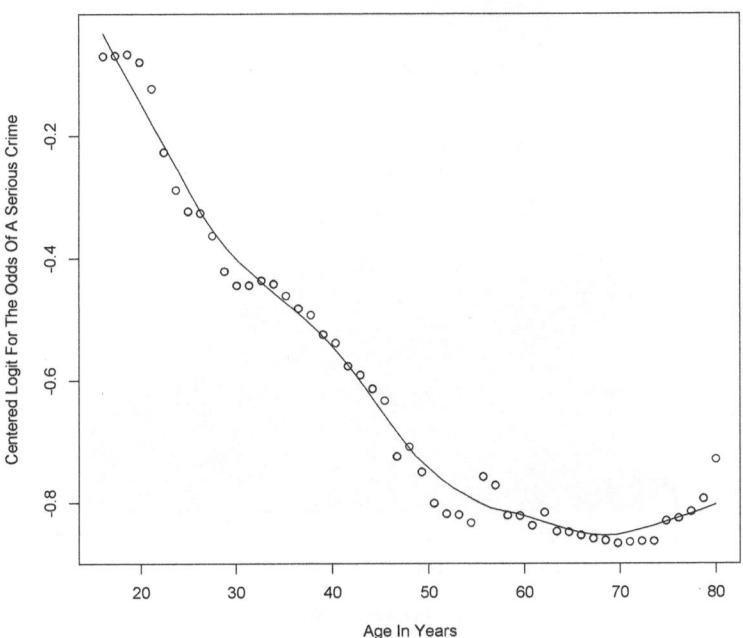

Fig. 6.8 Response plot of an arrest for a violent crime against years of age. The relationship is strongly negative until around age 50.

This effect is adjusted for the role of age — clearly, younger individuals are less likely to have committed crimes long ago, and age too is related to crime, as we will now see. Figure 6.8 shows the partial response plot for an arrest for a serious crime against age. The expected negative relationship materializes until around 50 years of

[15] There is no real data beyond a value of 50, but individuals with no prior serious offenses were coded as having a value of 100. A value of 100 is sufficiently long ago that the crime might as well never have occurred. This has no effect on the response function for values less than 50 because the fitted values are highly local. Researchers unfamiliar with local fitting methods, may find this surprising.

age, when the relationship becomes flat. In addition, it may be important in practice
to notice that the decline does not begin until around age 22. Individuals between 17
and 22 are about at the same high level of risk, other things equal. Once again, the
effect is large in a probability metric. Comparing an 18-year-old to a 50-year-old,
the probability of failure is about .22 less for the latter.

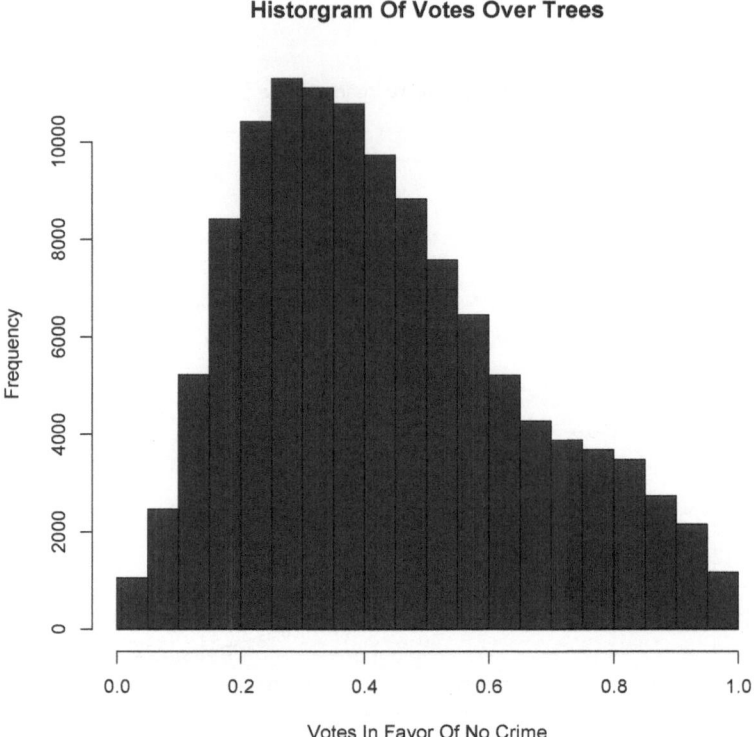

Fig. 6.9 A histogram of votes for no arrests over trees. The majority of those assigned to the no
arrests class are assigned by a substantial plurality. Overall uncertainty is relatively low.

The uncertainty issues are much the same as before. Random forests provides
vote tallies for each outcome class and for each observation. Ideally, the class assigned to an observation has a decisive percentage of the vote. The worst case is for
all of the percentages to be identical or nearly so. With three classes, for example,
three percentages of about 33% suggest a great deal of uncertainty in the assigned
outcome class. A 15%-35%-50% distribution, on the other hand, implies far less
uncertainty. But there is no line in the sand. Decision-makers should decide how decisive a vote needs to be for it to be decisive enough. As before, when the vote for a
given observation is not sufficiently decisive, it can be useful to exploit information

not available when the forecast was made to assist in decision-making. However, it is usually a mistake to use the same information to override a forecast, because forecasting accuracy is likely to be compromised.

Figure 6.9 shows the vote distribution for the response outcome of no arrests. One can see that a substantial number of those forecasted into the no arrests class received reasonably decisive vote. Those cases are forecasted with enough certainty so that decision-makers might well choose to proceed with no other information about likely post-release behavior. However, from the histogram alone it is difficult to get a very precise fix on how many such cases there are.

Here's why. A vote of 50%-25%-25$ is perhaps quite decisive. But what about a vote of 50%-48%-2%? The winning plurality is the same for both, but is the second really decisive? One probably needs to focus on the difference between the two highest vote percentages and decide that, say, a gap of at least 15 percentage points defines a sufficiently decisive vote. This can be easily done with the output from random forests but would require writing several lines of code.

When there are three response classes, there are graphical tools than can help. Figure 6.10 is a ternary plot showing for 50 cases the three vote proportions. In practice, forecasts to inform decisions will be made for a relatively small number of cases at a time and then, graphical displays can be very instructive. If the number of forecasts is large, "over-plotting" can make interpretation difficult.[16]

For Figure 6.10, the three response classes being forecasted are shown at the corners of the equilateral triangle. The plotting grid takes some getting used to, but is actually straightforward. Consider the single case closest to the lower corner of the triangle. For that individual, about 10% of the votes were for a violent crime, about 10% of the votes were for a nonviolent crime, and about 80% of the votes are for no crime at all. The vote is overwhelmingly for no arrests. The key to reading such plots is to recognize that the locations of any point are with respect to the vertical distance from a given side to the opposite triangle corner. For example, the vote proportions for no crime are represented by the vertical distance from the upper righthand side of the triangle to the corner labeled "none" on lower left side of the triangle. The proportions increase from 0.0 to 1.0 as one moves from the triangle side to the triangle corner. In short, ternary plots can be used to examine voting patterns on a case-by-case basis as decisions are being made.

Figure 6.10 can also be used to examine the patterns of votes for a set of forecasts as a whole. Ideally, there will be three clusters of points, each located close to one of the triangle corners. But a lot depends on how common an outcome class happens to be. If an outcome class is very rare, there can be no cluster. In this illustration, arrests for violent crimes are a distinct minority. With only 50 forecasts, one would not expect to find a large violent crimes cluster and in fact, there is at best a cluster of three a bit above .60. In contrast, there is a no-crimes cluster of nine cases with votes over 60% or better toward the lower left of the graph.

---

[16] The plot was constructed using the R procedure *ternaryplot* in the library *TeachingDemos*. Here's the code: *ternaryplot(votes, dimnames = c(V = "NonViolent",O = "None", V = "Violent"), main = "Ternary Plot for Class Votes",col = "blue", cex = .5, coordinates = T)*. The meaning each argument is well explained in the help documentation.

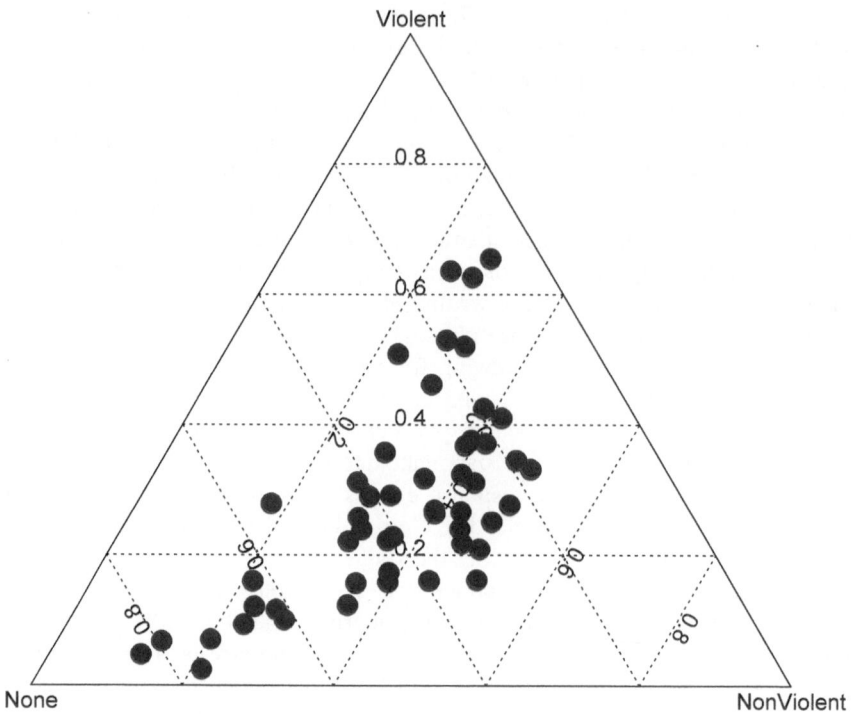

Fig. 6.10 A ternary plot of the distribution of votes for each of the three response classes. The proportions sum to 1.0.

In summary, Figure 6.10 indicates that for the 50 forecasts plotted, votes in favor of no arrests are more likely to be decisive than votes in favor of an arrest for a violent crime, and especially votes in favor of a nonviolent crime. There seems to be the most uncertainty about arrests for nonviolent crimes, which suggests that the class may be inherently ambiguous and the mix of crime etiologies more heterogeneous.

Chapter 7
Implementation

Abstract All of the material to this point would be at best academic if the key empirical results were unable to inform practice. This chapter turns briefly to porting new forecasting procedures to the settings in which they will be used. There are technical issues, but often the major obstacles are interpersonal and organizational.

7.1 Making it Real

We come finally to the matter of implementation. We now assume there exists an acceptable forecasting procedure that has been developed and evaluated. It is now time to port the forecasting procedure to the setting in which it can be used to inform real decisions.

Even more than during the developmental phase, implementation will require the talents and cooperation of many people. The range of expertise needed is much broader. The number of organizations likely to be involved is much larger. This is also the time when stakeholders will begin to fully appreciate the nature and consequences of the forecasts. As a result, they will engage more aggressively, and when they do, the media will not be far behind. It can all get quite lively.

7.2 Demonstration Phase

Starting with a demonstration phase is usually essential. Ideally, forecasting data of the sort that will be used later are tested. The goal is not to evaluate forecasting accuracy because that has already been done. Rather, the goal is to shake down the steps by which the forecasting data will be obtained and "dropped down" the random forests forecasting algorithm. For example, there needs to be information in the data carried along when forecasts are made than can link a given forecast to a given individual. This can mean that some official criminal justice ID is maintained throughout the process. Alternatively, if a new, unique ID is defined for each case, there is a way to go from the new ID to whatever unique identifiers are used in the official data bases.

Another issue can be whether the form the forecasts will take is easily understood by decisions makers. If the forecasts are to be provided in, for instance, a spread sheet format, accurate and easily understood labels for the rows and columns are essential. In some instances, all that decision-makers want is a color-coded forecasted category (e.g., red for serious crime, yellow for crime that is not serious, and green for no crime).

Any problems that surface during the demonstration phase are usually fixed quite easily. One should then repeat the demonstration because sometimes solutions to one problem create another problem. Only when all goes well should the actual implementation process begin.

7.3 Hardware and Software

The move to the physical environment in which the forecasting will be done raises hardware and software issues. Requirements include:

1. a computer running R, although in the near future other software may suffice;
2. an R object installed on that computer, containing random forests output constructed from the training data;
3. for that computer, direct access in real time to input data for forecasting, containing the values of predictors; and
4. an interactive user interface allowing users to specify the IDs of particular individuals and providing a convenient display of their forecasts.

No special computer is needed. Most personal computers have the processing power required. R is free and runs on a wide variety of platforms and under Windows, Mac OS, and Linux. If one assigns the output of random forests to some output name, and then saves that R object, that object will contain all that is required. It can be sent to another computer and saved there if that is desirable.

But problems can still occur. Perhaps most commonly, R output produced on one platform may not run properly on another platform. For example, it is very convenient in principle to port data, forecasts, and computer code in text format. Unfortunately, text file conventions for the Mac OS can be a little different from the text file conventions Windows. Likewise, data files created in an SPSS Windows format may create problems if read using Mac OS software. For example, dates can be scrambled. In short, there will likely be a few annoying details to be worked through.

7.4 Data Preparation for Forecasting

The data used for forecasting are usually treated as another random realization from the same joint distribution responsible for the training data. But working with new predictor values can be a little tricky. R is expecting that the predictors in the forecasting data have the *exact* same names as the predictors in the training data. For predictors that are *factors*, R is also expecting the same names for each *level*. For example, if in the training data the factor gender has male coded as "M" and female coded as "F", "m" and "f" in the forecasting data will not do. R is case sensitive. Nor will "male" and "female." One also has to plan ahead for working with codes for missing data (e.g., -99), blanks, and typos. For example, blanks can be represented by one space or many different numbers of spaces even in the same dataset.

It will likely be very difficult to anticipate every possible problem. So code must be written that will run nevertheless. Suppose there is a factor used as a predictor (a categorical variable) with the name X. Suppose in the training data X has three levels (i.e., categories): A, B, C. When random forests is applied, those three levels are automatically stored as an *attribute* of X. Random forests expects to see an A, B, or C whenever it processes X in the future.

Now suppose there is a set of new cases for which forecasts are needed. The predictor values for these cases constitute the forecasting data. Random forest output constructed for the training data provides the forecasting structure. The forecasting data are "dropped down" this structure so that forecasts can be obtained.

Suppose the predictor X is in both the training data and the forecasting data, but in the forecasting data the levels are A, C, and blank. This implies that no forecast can be made for all cases coded blank. The random forests forecasting object has no information on what to do with such cases. This would be true for any level found in the forecasting data but not in the training data (e.g., z, D, green, 17.5).[1] When there are any levels in the forecasting data not found in the training data, R will generate an error message saying as much.

There will also be no forecast for the level B because no cases in the forecasting data happen to have that level. There will be no error message, but one cannot expect to find any forecasts for cases that have a level B for variable X. This may well be disappointing, but it is not treated as an error by R. For example, perhaps the training data includes men and women, but the much smaller forecasting dataset happens to have only men. That is just the way it is.

In short, R allows for levels in the training data that are not in the forecasting data, but not for levels in the forecasting data that are not in the training data. The best solution for the latter is to "pre-process" the forecasting data so that the variable names and factor levels correspond to those in the training data. Unfortunately, errors can occur nevertheless because some data problems may slip through.

There needs to be a more reliable approach. When in R a factor is stored, attached is information on the variable's name, and that factor's legitimate levels. One must target the list of legitimate levels not the actual levels found in the data. The idea

---

[1] One cannot interpolate or extrapolate factors.

is to force the *recognized* levels for all factors in the forecasting data, whatever those levels actually are, to be the same as the *recognized* levels for the same factors in the training data. The code in R to make this happen can look something like this: levels(ForecastingData$V3) ← levels(TrainingData$V3). This code forces the levels for variable V3 in the forecasting data to be the same as the levels for variable V3 in the training data. No information is lost because a level for V3 that is not in the training data can have no empirical role in the forecasts.[2]

For *quantitative* variables, the data partitioning undertaken for each tree has no "holes"; there are no levels to wrestle with. For example, if a partition is defined for all individuals younger that 30, all ages under 30 are included even if not represented the training data. If there are no 26-year-olds in the training data, it does not matter. If there are no individuals under 19 in the training data, it does not matter. In short, quantitative variables are often easier to work with than qualitative variables.[3]

In summary, porting the forecasting R object to another platform should go relatively smoothly. Most of the effort will then go to getting routine access to the requisite predictors in real time and making sure that those data are pre-processed to be consistent with the training data. There may also be a need to write some computer code in case the pre-processing stumbles.

7.5 Personnel

The forecasting described in this book implies that even for technical matters, a team of individuals will be needed. There are just too many different skills required. It is probably unreasonable to expect that whoever is responsible for developing the forecasting procedures in R will also know the details of the various data bases required. Likewise, expertise with a particular agency's data processing platforms will depend on yet another kind of background. For this team to function, all of the key personnel need to buy into the project and be able play well with others. It is also essential that they be open with one another.

There are at least three major obstacles to assembling a good team. The first is a lack of expertise. Beyond usual problems finding a sufficient number of talented and motivated individuals, information technology is changing very rapidly. IT expertise should be date-stamped just like perishables in the grocery store. What worked last year may not work this year, and an advanced degree is no guarantee that the expertise is current.

---

[2] The function factor() can also be used to assign levels to a factor.

[3] The absence in the training data of certain values for a given predictor does not matter computationally for subsequent observations needing forecasts. But especially if the response function is highly nonlinear, the absence of certain values may mean that the function is not well-approximated. For example, should the shape of the function change dramatically for individuals under 19, and there are no such individuals in the training data, that change will not be captured. Then, when individuals under 19 are found in the forecasting data, their forecasts may be well of the mark.

The second is institutional rivalries. For example, a department of corrections may have longstanding tensions with a department of probation and parole. Yet, data may well be needed from both. It is rare to find explicit refusals to cooperate or outright sabotage. The most likely response is perfunctory agreement to cooperate followed by inactivity and silence. Often the only solution is a forceful kick-in-the-pants from one or more individuals sitting much higher up in the table of organizations.

The third is proprietary software owned by a for-profit organization. Proprietary software may be purchased easily enough, but important details may be inadequately disclosed, and performance claims may turn out to be unsubstantiated; there is usually no equivalent of peer review. It can also be difficult to alter the software, or require that changes be made by the purveyor, once the project has begun.

There is no recipe for how to overcome these obstacles. Perhaps the best advice is collect lots of information on the individuals and organizations who will likely be involved before deciding whether to proceed. If a decision is made to go forward, that information can also be used to help prevent problems before they occur. For example, if one is working with high level state agencies, getting in advance a sign-off on the project from the governor's office can help reinforce good behavior from all involved.

7.6 Public Relations

There can be at least three important public relations matters that are worth brief mention. First, it is very important from the start of the enterprise, well before implementation, that any statements about the forecasting should be careful, cautious, and tentative. Careful statements will convey no more than necessary and will be as accurate as the current understandings permit. Cautious statements will help prevent promising too much by, in part, communicating the key uncertainties of the project. Tentative statements allow for subsequent revisions, which will almost certainly be necessary.

Second, there needs to be a sensible division of labor. It can be helpful to have a single individual serve as the overall project spokesperson, with one additional, "specialized" spokesperson for each of the key components. For example, if requests are made for lots of details, there will need to be someone who can explain the forecasting procedures and someone who can explain how the forecasts will be used.

Third, there needs to be clear sign-off protocols for all written statements or reports. It is likely that any organizations with skin in the game will want to review all documents circulated to outsiders, and require that nothing be released without at least their knowledge and often, not without their permission as well. These requirements may need to be tailored for academic publications so that whenever possible, concerns about academic freedom do not arise.

Chapter 8
Some Concluding Observations About Actuarial Justice and More

Abstract This chapter offers some brief thoughts about what the future may hold for the procedures that figure centrally in the book, and discusses implications beyond the nuts-and-bolts problem of forecasting criminal behavior. A key topic is actuarial justice and several associated trends. There are also some brief observations about "black-box" forecasting tools, the future of criminal justice research, and something one might call "dark structure."

8.1 Current Trends in Actuarial Justice

To rephrase slightly a old aphorism, "information is power." As a matter of expediency and effectiveness, the information used in behavioral forecasting is increasingly quantitative — information is power and quantitative information can be especially powerful. Thus, investment strategies now depend heavily on "quants," which despite recent setbacks, represents an approach likely to accelerate. Mass marketing by the likes of Google, Amazon, and NetFlicks is relying more on enormous databases and less on Madison Avenue craft lore. Our national politics might grind to a halt were it not for the nearly-continual polling of prospective voters to project election outcomes. Political advisors now sound like survey analysts.

Similar trends are underway in criminal justice decision-making. In many jurisdictions, police deployments are informed by data systems like COMPSTAT (*Henry, 2003*), with more complete datasets coming on line regularly. One result is the growth of "predictive policing" in which traditional experience-based judgements can be either shored up or challenged by hard data. The routine use of data-driven forecasts to inform parole decisions has been spreading to sentencing, charging and bail recommendations. As with police, professional judgement can be reinforced or strongly challenged, and here too, vastly better data are becoming available. Data-intensive forecasting procedures are being built to inform decisions about juveniles, both as perpetrators and victims of violent crime.

Some warn that these trends contribute to the growth of "actuarial justice." Actuarial justice was perhaps first characterized over three decades ago in a essay reviewing a National Academy of Sciences Report on career criminals (*Messenger and Berk, 1987*). The essay addressed the use of forecasts to help identify individuals likely to have long and active criminal careers. A controversial option was to "incapacitate" with longer prison terms individuals projected to be "habitual offenders." The forecasts were based on offender profiles constructed from statistical models popular in that era. After considering a range of substantive and technical matters, the essay's authors offered several cautions.

Current practice is far more sophisticated and nuanced, but one can still reasonably ask whether actuarial justice is a good thing (*Feeley and Simon, 1994; Buruma, 2004; Robert , 2005; Monahan, 2006; Harcourt, 2007*). There is no doubt that actuarial methods can produce usefully accurate forecasts and improve cost-effectiveness across a wide variety of criminal justice agencies. But there are also legitimate concerns.

One concern is that cost-effectiveness is being allowed to shoulder aside other important criminal justice ideals. For instance, under "just deserts," sanction severity is supposed to be proportional to offense seriousness (*von Hirsh, 1995*). Future dangerousness has nothing to do with it. What about general deterrence in which sanctions applied to one individual are meant to dissuade others from committing crimes (*Farrell, 2003*)? Of necessity, one is drawn into the long-standing debate about the purposes of punishment (*Boonin, 2008*).

Another concern is that cost-effectiveness is, at least in practice, far too narrowly defined. Important "externalities" are neglected. Commonly mentioned, for example, are the consequences of incarceration for an offender's family. But, no conception of costs and benefits will cover every eventuality, and what can be credibly measured will always be a further subset. Perhaps the key question is whether a proposed collection of measurable benefits and costs is both feasible and a demonstrable improvement over current practice.

Yet another concern is actuarial reasoning itself. Hoping to make good decisions on the average is misguided. Rather, one should try to make the good decisions on a case-by-case basis. But criminal justice decision-makers usually make many decisions, some of which are surely better than others. A one-off framework seems to misrepresent the process, both in theory and practice. Should there not be some reasoned approach to good decisions in the aggregate?

Finally, there is the ongoing concern about the role of predictors, such race, ethnicity, and gender, that can be used for profiling. Here too, definitive answers are hard to come by. For example, David Boonin (*2011: Chapter 10*) makes an important distinction between "rational" profiling furthering some legitimate crime reduction goal and "irrational" profiling, which has no such impact. Under some circumstances, the former may be morally acceptable. There are also difficult technical matters. Among them is how the legal term "similarly situated" corresponds to statistical adjustments that "hold constant" a subset of covariates (*Berk, 2009*).[1]

---

[1] For example, when conventional regression analysis is used to empirically isolate the role of a suspect predictor such as race, covariance adjustments are required for all possible confounders.

Lying just behind each of these concerns is a fundamental policy matter: "Compared to what?" Actuarial justice has it problems, but so too do each of the alternatives. For example, why should one believe that discretionary decisions made by a parole board unguided by actuarial results will be more accurate, transparent and race-neutral? Or, if sentences are to be based on just deserts, how does one get from a broad, retributive right to punish to dose-like sentences that can range in length for several days to life (or death)? There may be a loose rank ordering of crime seriousness, but one requires far more than an ordinal scale (*Rossi and Berk, 1997*).

This is not the venue to examine these matters, and they are not be easily resolved in any case. But readers should be sensitive to the controversies that surround actuarial methods in criminal justice. Moreover, if an agency's use of actuarial methods is challenged from one or more critical perspectives, it is helpful to know that these perspectives can also be legitimately criticized. Some might say that currently, no one point of view holds the intellectual high ground.[2]

8.2 Larger Issues

Forecasts of human behavior are made all the time. Interpersonal interactions could not proceed without them. Major institutions are just as active. Admissions to law schools depend on projections of academic success, marketing strategies are guided by forecasts of how consumers will respond to a particular pitch, hiring choices anticipate how a new employee will work out, "best practices" recommended by various professional organizations are based expected results if those practices are adopted, and on and on. Criminal justice agencies are just one of many entities that routinely make and use forecasts. As such, they are part of several broad trends.

8.2.1 Forecasts Despite Individual Uniqueness

No two people are exactly alike. If taken at face value, forecasts of individual behavior are effectively impossible. One cannot learn anything from training data that can be usefully applied to the forecasting data.[3]

This is an impossible standard to meet. In practice, researchers will settle for all of the "important" confounders. How close to the impossible standard does one have to be in order to claim that cases are similarly situated? And how would one know? In addition, as a mathematical matter, the racial parameter whose value is being estimated differs depending on the confounders taken into account; there is not one potential race effect, but many. The issues can be subtle.

[2] If it is any consolation, similar issues come up in discussions of heath care, environmental protection, and national defense.

[3] A given individual is not identical from moment to moment. Even within-person forecasts might seem problematic. But there is no need to go there for now.

But, there is ample evidence that useful forecasts are being made all the time in a wide variety of settings. In practice, there are likely to be a least several individuals in training and forecasting data who are *sufficiently alike* to make forecasting possible. With very large data bases, "several" can become "many." At that point, the statistical tools discussed earlier can be introduced. A set of similarly situated individuals becomes a profile. In short, there is no necessary contradiction between each individual's uniqueness and useful forecasts of individual behavior.

8.2.2 Risk and Forecasting

The current discussions about health care and the role of government more generally have raised in a very visible fashion the need to consider risk. One evaluates whether the dollars spend to manage some undesirable outcome are worth the money. As part of that discussion, risk is properly unpacked so that the probability of an event and the consequences of that event can be addressed separately. This is just as it should be if decisions are to be better informed. Closely related considerations are now quite common across a wide variety of criminal justice decisions, and may soon be nearly universal. Perhaps the key take-home message is that with a growing appreciation of risk comes a growing appreciation of forecasting. Risk is, after all, about the future.

8.2.3 Black-Box Methods

Actuarial methods are changing rapidly. Forecasts increasingly exploit enormous datasets that are routinely available in real time. Google, for instance, builds customer profiles on hundreds of thousands of clicks. Major supermarket chains build customer profiles from millions of in-store transactions. Criminal justice forecasting is moving in a similar fashion to very large datasets. Administrative datasets with over 100,000 observations are common. All of these applications share a common insight: there is substantial predictive structure to be found in very small subsets of observations, which ordinarily would be ignored, but in the aggregate can improve forecasting enormously. It takes very large datasets to exploit such structure. As observed earlier, if you are looking for a needle in a haystack, you first need a haystack.

To capitalize on information in very large datasets, especially if that information can be spread over many small partitions of the data, one must apply appropriate statistical procedures. Often this means machine learning, since conventional regression approaches are likely to stumble. But there can be important tradeoffs.

One salient feature of machine learning is that it makes black box methods a virtue (*Breiman, 2001b*). The goal is to link "inputs" to "outputs" so that the outputs correspond as closely as possible to the observed outcomes. Forecasting then follows naturally. But *how* inputs are related to outputs is a secondary concern. Indeed, in some circumstances, the predictors can be viewed primarily as providing a vehicle for undertaking an effective and efficient search of the data. Their meaning is far less important than their Sherpa-like capacity to guide a machine learning algorithm through a high-dimensional predictor space. One result is that forecasting becomes a distinct activity that differs from explanation or conventional "risk assessment." What matters most is forecasting accuracy. Combining explanation with forecasting can compromise both.

A related result is that the underlying statistical formalization will be very different from statistical models popular in the social sciences. There is no causal inference because there is no causal model. In exchange, the data generation model is typically far less demanding of subject matter knowledge, and one is free to proceed in a highly inductive manner.

Finally, good forecasting practice should incorporate information on how the forecasts will be used as well as the consequences of forecasting errors. This fits well within machine learning perspectives, but also forces a marriage of policy and statistics. Practitioners of either will perhaps squirm, at least a bit. Some stakeholders will argue that this is a good thing.

8.3 Dark Structure and New Science

At the risk of some hyperbole, the machine learning approach to very large datasets has the prospect of creating new science. As we saw earlier, there can be forecasting skill beyond that usually obtained from conventional regression. A natural question is how this can be; by and large, the predictors used in a machine learning approach are the same as those used in conventional regression. Somehow, more predictive information is being extracted.

One reason is that the inductive approach is effectively nonparametric. No functional forms need to be imposed *a priori*. Recall that highly nonlinear relationships and high-order interactions can be unearthed. In criminal justice applications at least, many such relationships have been found that are effectively unforeseen by criminology theory and the usual parametric regressions. Both will have some catching up to do.

Another reason is that the nonlinear relationships and high-order interactions can provide, as explained earlier, surrogates for predictors that are not explicitly included. These may go well beyond the usual suspects. Determining what these predictors might be will take some careful detective work, but there is promise. What are the sources of associations that go beyond a list of well-known predictors?

A final reason is that when very large datasets are coupled with machine learning, one can find many small regions of the predictor space where a few observations are related to an outcome in a systematic fashion. Such structure is ordinarily ignored because any effects are too small to be important. But as now said several times, there is important predictive information when one considers many such regions at once.

These three sources of predictive information imply that there is substantial structure in criminal behavior that current thinking in criminal justice circles does not consider. One might use the term "dark structure" because so little is known about its characteristics. We know it is real because it improves prediction. But it remains somewhat mysterious. For the next generation of criminal justice researchers, this is surely good news.

8.4 A Closing Sermonette

Forecasts of criminal behavior will often get it wrong. There are just too many factors involved within a highly nonlinear system. The proper benchmark, therefore should not be perfection. The proper benchmark is current practice. And in that context, accuracy is not the only goal. Also important are fairness and transparency. Although some may bristle at the idea, criminal justice agencies are substantially engaged in risk management. Better forecasts surely can help.

References

1. Baldi, P., & Brunak, S. (2001) *Bioinfomatics: The Machine Learning Approach*. second edition, Boston: MIT Press
2. Berk, R. A. (2008a) Forecasting methods in crime and justice. In J. Hagan, K. L. Schepple, and T. R. Tyler (eds.) *Annual Review of Law and Social Science* 4 (173–192). Palo Alto: Annual Reviews.
3. Berk., R. A. (2008b) *Statistical Learning from a Regression Perspective*. New York: Springer.
4. Berk., R. A.(2009) The role of race in forecasts of violent crime. *Race and Social Problems* 1(4): 231–242.
5. Berk, R. A., Brown, L., & Zhao, L. (2010) Statistical inference after model selection. *Journal of Quantitative Criminology*, 26(2): 217–236.
6. Berk. R. A., Kriegler, B., & Baek, J-H. (2006) Forecasting dangerous inmate misconduct: An application of ensemble statistical procedures. *Journal of Quantitative Criminology* 22(2) 135–145.
7. Berk, R. A., Sorenson, S. B., & He, Y. (2005) Developing a practical forecasting screener for domestic violence incidents. *Evaluation Review* 29(4): 358–382.
8. Berk, R. A., Sherman, L., Barnes, G., Kurtz, E., & Ahlman, L. (2009) Forecasting murder within a population of probationers and parolees: A high stakes application of statistical learning. *Journal of the Royal Statistics Society — Series A* 172 (part I): 191–211.
9. Boonin, D. (2011) *Should Race Matter?* Cambridge: Cambridge University Press.
10. Boonin, D. (2008) *The Problem of Punishment*. Cambridge: Cambridge University Press.
11. Blumstein, A., & Nakamura, K. Redemption in the presence of widespread criminal background checks. *Criminology* 47(2): 327–359.
12. Borden, H. G. (1928) Factors predicting parole success. *Journal of the American Institute of Criminal Law and Criminology* 19: 328–336.
13. Breiman, L. (1996) Bagging predictors. *Machine Learning* 26:123–140.
14. Breiman, L. (2001a) Random forests. *Machine Learning* 45: 5–32.
15. Breiman, L. (2001b) Statistical modeling: two cultures (with discussion). *Statistical Science* 16: 199–231.
16. Breiman, L., Friedman, J.H., Olshen, R.A., & Stone, C.J. (1984) *Classification and Regression Trees*. Monterey, CA: Wadsworth Press.
17. Buruma, Y. (2004) Risk assessment and criminal law: closing the gap between criminal law and and criminology (pp. 41–61). In G. Bruinsma and H. Elffers (eds.) *Punishment, Place, and Perpetrators: Developments in Criminology and Criminal Justice Research*. Portland, OR: Willan Publishing.

18. Burgess, E. M. (1928) Factors determining success or failure on parole. In A. A. Bruce, A. J. Harno, E. .W Burgess, and E. W., Landesco (eds.) *The Working of the Indeterminate Sentence Law and the Parole System in Illinois* (pp. 205–249). Springfield, Illinois, State Board of Parole.
19. Casey, P. M., Warren, R. K., & Elek, J. K. (2011) Using offender risk and needs assessment information at sentencing: guidance from a national working group. National Center for State Courts, www.ncsconline.org/.
20. Chipman, H. A., George, E. I., & McCulloch, R. E. (1998) Bayesian CART model search (with discussion). *Journal of the American Statistical Association* 93: 935–960.
21. Chipman, H. A., George, E. I., & McCulloch, R. E. (2010) BART: Bayesian additive regression trees. *Annals of Applied Statistics* 4(1): 266–298.
22. Culp, M., Johnson, K., & Michailidis, G. (2006) ada: an R package for stochastic boosting. *Journal of Statistical Software* 17(2): 1–27
23. Dean, C. W., & Dugan, T. J. (1968) Problems in parole prediction: a historical analysis. *Social Problems* 15: 450–459.
24. Efron, B., & Tibshirani, R.J. (1993) *An Introduction to the Bootstrap* London: Chapman & Hall.
25. Farrell, D. (2003) The justification for general deterrence. In D. Matravers, and J. Pike (eds.) *Debates in Contemporary Political Philosophy: An Anthology*. New York: Routledge.
26. Farrington, D. P. & Tarling, R. (2003) *Prediction in Criminology*. Albany: SUNY Press.
27. Feeley, M., & Simon, J. (1994). Actuarial justice: The emerging new criminal law. In D. Nelken (ed.), *The Futures of Criminology* (pp. 173201). London: Sage Publications.
28. Friedman, J. H. (2002) Stochastic gradient boosting. *Computational Statistics and Data Analysis* 38: 367–378.
29. Glaser, D. (1955) Testing correctional decisions. *The Journal of Criminal Law, Criminology and Police Science* 45: 679–684.
30. Gottfredson, S. D., & Moriarty, L. J. (2006) Statistical risk assessment: old problems and new applications. *Crime & Delinquency* 52(1): 178–200.
31. Harcourt, B.W. (2007) *Against Prediction: Profiling, Policing, and Punishing in an Actuarial Age*. Chicago, University of Chicago Press.
32. Hastie, R., & Dawes, R. M. (2001) *Rational Choice in an Uncertain World*. Thousand Oaks: Sage Publications.
33. Hastie, T. & Tibshirani, R. (2000) Bayesian backfitting. *Statistical Science* 15(3): 196–223.
34. Hastie, T., Tibshirani, R., & Friedman, J. (2009) *The Elements of Statistical Learning*. Second Edition. New York: Springer.
35. Heller, K. (2012) Karen Heller: Philadelphia's murder rate is a deadly, costly epidemic. *Philadelphia Inquirer* January 4, 2012.
36. Henry, V.E. (2003) *The Compstat Paradigm*. New York: Loose Leaf Law Publications.
37. Ho, T.K. (1998) The random subspace method for constructing decision trees. *IEEE Transactions on Pattern Recognizition and Machine Intelligence* 20 (8) 832–844.
38. Hothorn, T., Buehlmann, P., Kneib, T., Schmidt, M., & Hofner, B. (2011) Package "mboost". cran.r-project.org/web/packages/mboost/mboost.pdf.
39. Hyatt, J.M., Chanenson, L. & Bergstrom, M.H. (2011) Reform in motion: the promise and profiles of incorporating risk assessments and cost-benefit analysis into Pennsylvania Sentencing. *Duquesne Law Review* 49(4): 707–749.
40. Kleiman, M., Ostrom, B. J., & Cheeman, F. L. (2007) Using risk assessment to inform sentencing decisions for nonviolent offenders in Virginia. *Crime & Delinquency* 53(1): 1–27.
41. Leeb, H., & Pötscher, B.M. (2005) Model selection and inference: facts and fiction," *Econometric Theory*21: 21–59.
42. Leeb, H., & Pötscher, B.M. (2006) Can one estimate the conditional distribution of post-model-selection estimators? *The Annals of Statistics* 34(5): 2554–2591.
43. Liew, T.W. (2011) California prisons: "Non-revocable parole" is too dangerous. *Los Angeles Times*, OP-ED, June 11, 2011.
44. Mease, D., Wyner, A.J., & Buja, A. (2007) Boosted classification trees and class probability/quantile estimation. *Journal of Machine Learning Resaerch* 8: 409–439.

45. Messinger, S.L., & Berk, R.A. (1987) Dangerous people: a review of the NAS report on career criminals. *Criminology* 25(3): 767–781
46. Monahan, J. (1981) *Predicting Violent Behavior: An Assessment of Clinical Techniques*. Newbury Park: Sage Publications.
47. Monahan, J. (2006) A jurisprudence of risk assessment: forecasting harm among prisoners, predators, and patients. *Virginia Law Review* 92: 391–435.
48. Monahan, J., & Solver, E. (2003) Judicial decision thresholds for violence risk management. *International Journal of Forensice Mental Health* 2(1): 1–6.
49. Ohlin, L. E., & Duncan, O. D. (1949) The efficiency of prediction in criminology. *American Journal of Sociology* 54: 441–452.
50. Ohlin, L. E., & Lawrence, R. A. (1952) A comparison of alternative methods of parole prediction. *American Sociological Review* 17: 268–274.
51. Pew Center of the States, Public Safety Performance Project (2001) Risk/needs assessment 101: science reveals new tools to manage offenders. The Pew Center of the States. www.pewcenteronthestates.org/publicsafety.
52. Reiss, A. J. (1951) The accuracy, efficiency, and validity of a prediction instrument. *American Journal of Sociology* 17: 268–274.
53. Rice, J. A. (2006) *Mathematical Statistics and Data Analysis*. Third Edition. New York: Duxbury Press,
54. Ridgeway, G. (2007) Generalized boosted models: a guide to the gbm package. cran.r-project.org/web/packages/gbm/vignettes/gbm.pdf.
55. Robert, D (2005) Actuarial justice. In M. Bosworth (ed.) *Encyclopedia of Prisons and Correctional Facilities* Vol 1, pp. 11–14. Thousand Oaks, CA: Sage Publications.
56. Rossi, P.H. & Berk, R.A., (1997) *Just Punishment: An Empirical Study of the Federal Sentencing Guidelines*. New York: Adline de Gruyter.
57. Skeem, J. .L., & Monahan, J. (2011) Current directions in violence risk assessment. *Current Directions in Psychological Science* 21(1): 38–42.
58. Sorensen, J. R., & Pilgrim, R. L. (2000) An actuarial risk assessment of violence posed by capital murder defendants. *The Journal of Criminal Law and Criminology* 90: 1251–1270.
59. von Hirsh, A. (1995) *Censure and Sanctions*. Oxford: Oxford University press.
60. Wilkins, L. T. (1980) Problems with existing prediction studies and future research needs. *The Journal of Criminal Law and Criminology* 71: 98–101.